W9-ABN-303

MELVILLE MANIFESTOS

DIVINE
DESTRUCTION

DIVINE
DESTRUCTION

WISE USE, DOMINION THEOLOGY
AND THE MAKING OF AMERICAN
ENVIRONMENTAL POLICY

STEPHENIE HENDRICKS

MELVILLE HOUSE PUBLISHING
HOBOKEN, NEW JERSEY

24.95

BOOK DESIGN: DAVID KONOPKA

MELVILLE HOUSE PUBLISHING
P.O. BOX 3278
HOBOKEN, NJ 07030

FIRST MELVILLE HOUSE PRINTING
ISBN: 0-9766583-4-8

PRINTED IN CANADA

CONTENTS

DIVINE DESTRUCTION

"And God blessed them, and God said unto them, Be fruitful, and multiply, and replenish the earth, and subdue it: and have dominion over the fish of the sea, and over the fowl of the air, and over every living thing that moveth upon the earth."

—GENESIS 1:28

"Defile not therefore the land which ye shall inhabit."

—NUMBERS 35:34

THUGS IN
THE WOODS

When I first encountered Judith Spencer, in June 2003, she should have been happier.

In 1999, after a long, demanding career as a nurse in San Jose, California, she and her husband Bob had decided to retire and leave behind the noise and pollution of crowded San Jose to settle permanently into what had been, since 1980, their vacation home—a small, comfortable place in the picturesque Sierra Nevada Mountains of northern California, right on the edge of the Stanislaus National Forest in the town of Arnold, California. It was remote, quiet, and beautiful—a dream location for a retirement home, Judith believed, and for attempting to realize another dream she had, that of becoming a writer.

I'd been put in touch with Judith by a contact that thought she could help me with some research I was *conducting. As a journalist and an outdoorswoman, I'd been hired to conduct a report for a forest preserva-tion organization about threats to the national forests of the Sierra Nevada. Judith could give me an idea of what was going on in the Stanislaus, said my contact. And that she did. It wasn't, however, what I expected.*

Judith, as it turned out, had in short order discovered that she had not moved into the perfect writer's refuge.

For one thing, living on the fringe of a national forest, she and Bob soon learned, now means living with the roar of engines. It seemed the federal wilder-ness has become more populated by various kinds of off-road vehicles (known collectively as ORV's) than by animals. The forest is now populated by motorbikes, by three-wheel all-terrain vehicles, by four-wheel-drive jeeps and trucks, by SUV's and Hummers. They were all over the Stanislaus National Forest—and most other national forests, I would come to find out—causing erosion, destroying fragile streams and wetland ecosystems, and leaving behind garbage, pollution, and waste as they went.

Making things worse, the people who enjoyed tearing through the wilderness on loud, gas-guzzling vehicles were often doing so on extra fuel themselves. According to Judith and her neighbors, not to mention park rangers and local police reports, wild weekends with the ORV-gang were often fueled by beer, at the very least, and hard drugs at the worst. Meanwhile, the sound of gunfire every now and then did nothing to make the neighbors feel more at ease. And what really terrified locals was the most widely used drug of them all: cigarettes, the cause of an increasing number of lethal forest fires in California. The fear of rowdy tourists was one thing. The fear of catastrophic forest fires was another.

It was not, as I say, what I expected to find. Nor would anyone have expected, I think, the extremes to which the situation had gone.

Judith and Bob Spencer were not the only ones disturbed by what was going on in the Stanislaus. Others in the community—which had seen a new influx of people from the San Francisco area escaping to the safety of the mountains in the wake of the 9/11 attacks—were also upset to find their idyllic setting so despoiled.

They organized themselves to do what Americans can do in such situations: lobby local official to enact legislation to protect the community on their side of the forest and prompt the federal government to better control the situation on its side. Essentially, Judith and her neighbors met in each other's kitchens trying to figure out how to make their voices be heard. They were particularly interested in stopping a development planned for an area near the edge of the Stanislaus near Arnold: Trees were to be cut down to make way for an ORV track. They called their group Commitment to Our Recreational Environment—CORE. It was an ad hoc organization, but Judith was the one usually elected to speak at town meetings, and she was the one often enough portrayed in the local newspaper as CORE's leader.

And so it was that Judith was the one who got the death threats.

One was a crude drawing, showing her being torn apart by all-terrain vehicles—ATV's—going in opposite directions; it was an American update of the execution dealt out to French and English traitors until the eighteenth century.

Another was a simple note, saying her days were numbered, in the most vulgar language imaginable.

Ominous postings and flyers appeared in mailboxes and on telephone polls throughout Arnold.

The local sheriff was understaffed and ill equipped to control the growing number of "ORVers"—as the users of ORV's call themselves—flooding the community every weekend and holiday, let alone conduct an investigation to track down whoever was writing the death threats. Meanwhile, the sheriff got little help from the rangers stationed in the Stanislaus. They weren't policemen, and after all federal laws and regulations permitted ORV activity. In fact, everything about U.S.

policy these days is encouraging even more development in the National Park system to accommodate such tourism. For the Spensers and many others in similar situations on the borderlands of what should be paradise, ORV users are posing an even bigger threat to forests than the timber interests that environmentalists used to be most concerned about.

However, this story doesn't end with death threats, for despite the thuggish, illegal, and offensive ways the ORVers threatened Judith's life, they responded to CORE's work, we might say, with their own political answer. The ORVers believed that no citizen should be able to interfere with the recreational activities they enjoyed. Their threats were a visceral vocalization of this belief.

But the ORVers were not simply fighting for their right to ride in and around the Stanislaus National Forest as they pleased. As it turned out, many ORVers believed that they had a much more substantial right to use the forest. Indeed, many riders made an ideological claim: They understood, as I would come to find out over the next year, that the national forests were to be used, no matter the implications, for private gain— including, but not limited to, their own recreational activities. Their thinking about the issue was almost directly opposite from Judith and her fellow organizers. Instead of wanting to preserve the forest for future generations, many ORVers thought that a public forest was open to any group who could find a real use for it, especially one as dramatic and popular as their own. The ORVers were also well organized. Many were active members of groups, clubs, and societies organized around their principles.

More troubling, however, was the fact that some ORVers near Stanislaus found justification for their activities in religion. Many of the riders were active in fundamentalist Christian groups who believed, as outrageous as it may seem, that their activities were sanctioned by God. A preoccupation with recreational vehicles seemed to allow riders, near Stanislaus and elsewhere, to interpret various passages of the Bible in peculiar ways. For example, the Colorado Springs Christian Four Wheelers, whose website I would find much later, shows photos of various members of their group tearing up mountainsides on all-terrain-vehicles and features a quote from Micah 4:2, "Come let us go up on the mountains"—an invocation not unlike the ones used by riders near Stanislaus.

Most peculiar among the many interpretations of the bible that I discovered—one that is shared by others I would come to find later—was one that proclaims that the Second Coming of Christ, and the ascent of all Christians into heaven, hinges on the exhaustion of our natural resources. It is a belief that has a complicated relationship to the Bible, for it requires that one believe that God's call to have dominion over the earth is taken literally—but one must also feel that the "End of Times" spoken about in *The Book of Revelations* is near. Some who concur with such an interpretation believe that global environmental annihilation is a divine requirement for Christ's return. It's an idea so foreign to the environmental movement that it requires pause to even comprehend.

I'd gone to the Stanislaus expecting something else, the typical situation in the Northwest that I'd covered before: I'd expected to find myself document-ing timber industry incursions on the national forests of the Sierra Nevada. Instead, I'd learned that Judith, a pleasant senior citizen who had simply asked some Mad Max wannabes in the woods to keep it down, had had her life threatened. But I had also discovered a large, well-organized group of religious ideologues, people who were politically active and especially vocal.

What I realized very quickly was that I had uncovered a far bigger story, for many of the organi-zations that the Stanislaus ORVers belonged to had allegiances with larger and better-funded advocacy organizations, many of which had direct ties to the Bush administration. Of course, ties between the administration and the New Christian Right have *already been well documented. It now became clear, however, that American environmental policy was also being greatly influenced, even shaped, by certain Christian Fundamentalists. Indeed, anti-environmental ideologues and members of the Christian Right share a thinly veiled consensus of ideology that implies a divine imperative to their policies.*

The evidence of this seemed substantial and wide-ranging, and yet it is little known to the general public. As with much of the mainstream discussion of contemporary American politics, it is as if those who know about it can't believe it, or are intimidated, somehow, into not talking about. For as I was also to learn, at the suggestion that an extreme religious ideology may be involved in the creation of American environmental policy, most people—even environmental activists—invariably fall into an uncomfortable silence.

CHAPTER I

THE PERVERSION OF "WISE USE"

In 1845, New York journalist John O'Sullivan editorialized that "it was the nation's *manifest destiny* to overspread and to possess the whole of the continent which Providence has given." With this, O'Sullivan coined the phrase "Manifest Destiny," an expression that would remain short-hand during the late nineteenth century for the belief that Americans had an obligation to settle the western territories. Indeed, the phrase "Manifest Destiny" implied that America's expansion was predetermined, undeniable, and—most importantly—inspired by God.

The ideas that precipitated talk about and belief in Manifest Destiny, however, were not necessarily the most important cause for America's population expansion in the west. Rather, throughout the late

1840s, Manifest Destiny was taken up and used as a rallying cry by those in government who had *already* wanted the entirety of the North American continent settled. A religious rationale that echoed this aim was a convenient and useful way to ask Americans to go west. Quite simply, arguments about Manifest Destiny provided only the rationale for Westward expansion, not its impetus.

Today, large, well-organized, and powerful groups of anti-environmental activists are using similar tactics. The anti-environmental philosophy known as "Wise Use" has gained a large audience, and many of its advocates and thinkers hold a menacing influence over government. A frightening fact in its own right, the widespread acceptance of anti-environmental thinking in the guise of Wise Use is made more troubling in that there are increasingly close ties between those who subscribe to the ideas of Wise Use and members of fundamentalist Christian churches and organizations. The Wise Use movement's influence over religious conservatives thus mirrors a traditional relationship between religious and political conservatives in that Wise Use advocates are increasingly adapting their own agenda to include the concerns of religious voters. In so doing, they have gained an army of God to promote their own agenda.

Although many credit the modern day right-wing activist (and timber industry consultant) Ron Arnold with having coined the term "Wise Use," the phrase actually originated a century before with the man appointed head of the U.S. Forest Service by Theodore Roosevelt, Gifford Pinchot. He used the term in his 1903 book *A Primer of Forestry*, partly in response to intense pressure from railroad companies to use Forest Service lands.* Pinchot believed that a balance should be struck between caring for the forests and man's interests.

But Pinchot also formulated the term in response to the views of famous naturalist John Muir, another Roosevelt associate, who advocated that public lands remain completely untouched. Thus, Pinchot's position was that of a somewhat enlightened government official needing to find a compromise between business interests, the concerns of naturalists, and the still ongoing expansion of the population into the western territories. He was trying, in short, to find a way to protect the emerging concept of public land from what Roosevelt called the "land grabbers." Wise Use, as it was originally conceived, allowed the western

---

* Pinchot wrote: "...the fundamental idea in forestry is that of perpetuation by wise use; that is, of making the forest yield the best service possible at the present in such a way that its usefulness in the future will not be diminished, but rather increased.... There is grave danger that the best of our forests will all be gone before their protection and perpetuation by wise use can be begun."

territories to prosper, and continue doing so, while preserving many forests and other natural environments for future generations.

While Ron Arnold did not originally coin the term Wise Use—he maintains, however, that he has coined other terms such as "ecoterrorist" and "rural cleansing"—he has come a long way in redefining the concept from the way it was initially used by Gifford Pinchot. Indeed, Arnold is generally considered the "father" of this modern-day incarnation of Wise Use, and he is particularly well-known for a series of sophisticated writings about the environment in which he has, since the mid-1980s, conceptualized a combative critique of the environmental movement that is deeply ideological. He is currently executive vice-president of a think tank that, although a non-profit, un-ironically calls itself The Center for the Defense of Free Enterprise (CDFE) and which, according to its website, monitors "threats to free markets, property rights and limited government." It also serves as a key center for anti-environmental activism. Vice President Dick Cheney is on the CDFE's board of directors.

In a recent attempt to clearly define Wise Use, Arnold critiqued the central tenets of the environmental movement, claiming that concealed below its talk of conservation was a radical political agenda, one that he believed aimed to "hamper property rights" and "dislodge the market system with public ownership of all resources and production." Arnold argues that, since the 1980s the environmental movement has moved into the mainstream and become "the Establishment;" he describes his own Wise Use movement as a "competing paradigm" to the environmental movement as he understands it. Arnold proclaims that the solutions to the world's environmental problems, whatever they may be, will be found by leaders in technology, industry, and trade—not by the environmental movement, as he believes is widely assumed. In short, Arnold declares that we need natural resources to survive and prosper and can survive any side effects of their use. *Our limitless imaginations can break through natural limits to make earthly goods and carrying capacity virtually infinite*," he writes.[1]

Lobbying by Wise Use groups has quickly precipitated the government to sanction various timber industry exploitations of public forests (often, as I have reported in earlier stories, with very little compensation to the U.S. taxpayer), the development of resorts in our national parks, and the opening up of Forest Service land to off-road vehicle use. The accommodation of ORVers, in fact, may represent the most profitable exploitation of publicly held land yet, especially considering that the private companies that stand to benefit from increased ORV access to the parks and forests—chiefly, the world's automotive manufactures—are some of the biggest corporations in the world.* The history of these environmental concessions, particularly the transfer of public land to private interests, shows the efficiency and power of the Wise Use movement's advocacy work.

While Arnold's aggressive conceptualization of the Wise Use movement was not codified until the publication of his 1989 book, *The Wise Use Agenda*, as early as 1979 he was giving the logging industry advice that revealed his innovative tactical thinking in its early stages. In a series of articles for *Logging Management Magazine*, he wrote:

> Citizen activist groups, allied to the forest industry, are vital to our future survival. They can speak for us in the public interest where we ourselves cannot. They are

---

* General Motors, Suzuki, Yamaha, Ford, and even Toyota, an early developer of environmentally friendly hybrids, all seem to be working hard to foster the idea that America's wilderness should be opened up to off-road vehicles. To name only one example, all sponsor television commercials showing wholesome families marveling at the way their SUV is tearing up the pristine wilderness. (And, as I would eventually discover, many also buy advertising on even the most amateurish ORV club websites.) And it seems logical to assume that further development is on their minds when these companies make their political donations, as well; General Motors was one of the largest contributors to George Bush's last presidential campaign, for instance.

not limited by liability, contract law or ethical codes...
industry must come to support citizen activist groups,
providing funds, materials, transportation, and most of
all, hard facts.[2]

A few years later, while addressing representatives of the Canadian logging industry, he bluntly restated these tactics that he would go on to develop to great effect: "Give them [the pro-industry groups] the money. You stop defending yourselves, let them do it, and you get the hell out of the way. Because citizen's groups have credibility and industries don't."[3]

David Helvarg, perhaps the leading authority on Arnold and the Wise Use movement, has detailed in numerous reports from the field—and particularly in his landmark book, *The War Against the Greens*, that the modern Wise Use movement really kicked into gear in the late 1980s, in response, says Helvarg, to "the perceived threat that George H. W. Bush would follow through on his pledge to be 'the environmental president.'"[4]

Helvarg notes that in the battle against groups such as the Nature Conservancy and the Sierra Club, Wise Use activists not only used Arnold's idea of forming fraudulent citizen activist groups that were actually funded by industry, but they also used "vigilante-style tactics ranging from telephone death-threats to arson and shootings."[5] According to Helvarg, "In Washington, Idaho, Montana, and New Mexico, a number of wise-users even united with the militia movement." But it was a tactic that backfired, Helvarg thinks, after the 1995 attack on the Murrah Federal Building in Oklahoma City by militia associates Timothy McVeigh and Terry Nichols, when much of the movement's industry funders backed away.[6]

Still, despite the tactical and political setbacks experienced by the Wise Use movement during the Clinton administration, the current movement has been reinvigorated. In particular, it has reached new constituencies and found a sympathetic executive. The recovery came about largely under the guidance of Arnold (who quickly tempered his habit of inflammatory language, such as when he said "We want to destroy environmentalists," and that he wanted to "kill the bastards").[7] Also important to that recovery was the leadership of the man who had been Ronald Reagan's "notoriously anti-environmental Interior secretary," James Watt (who, after being fired by Reagan,* told a group of ranchers that "if the troubles from environmentalists cannot be solved in the jury box or at the ballot box, perhaps the cartridge box should be used.")[8] Many current and recent government officials who oversee the environment, such as George W. Bush's current Interior Secretary Gale Norton, and Bush's former Agriculture Secretary Ann Veneman, were protégés of Watt, and worked for him at his Mountain States Legal Foundation, which has billed itself as the "litigation arm of Wise Use."

But what may have contributed most to that recovery is the connections the Wise Use movement has made with the New Christian Right. In a strategy seemingly modeled on Ron Arnold's directive to "stop defending yourselves" and "get the hell out of the way," some leaders in the Wise Use movement reached out to collaborate, especially in the late 1980s, with the newly re-politicized Christian Right. As this collaboration has become more substantial, Wise Use activists have had increasing access to local, regional, and national Christian grassroots organizations. Such collaboration between religious and political ideologues recalls, in many ways, the mobilization of Manifest Destiny more than 150 years earlier.

---

* Watt was fired in 1983 after he responded to criticism of his authorization of the sale of more than a billion tons of coal from Federal lands by saying that his coal-advisory panel included "a black, a woman, two Jews, and a cripple." In 1990 he pled guilty to charges of withholding documents from a Federal grand jury investigating a Reagan-era Department of Housing and Urban development scandal. In return, he was sentenced to five years' probation and 18 felony perjury charges against him were dropped.

A clear example of Wise Use's historic influence over the New Christian Right can be found by looking at the peculiar evolution of the Oregon Citizens Alliance (OCA), a group founded in the ashes of the 1986 electoral defeat of Joe P. Lutz. Lutz was a fundamentalist minister trying to wrest the Republican nomination for Senator away from incumbent Bob Packwood, whose moderate and pro-choice stance angered many fundamentalists; Lutz barely lost, garnering 42 percent of the vote.

Much like the impressive conservative mobilization after the presidential defeat of Barry Goldwater and the conservative Christian mobilization after the presidential defeat of Pat Robertson, Lutz's supporters created OCA as an ultra-right-wing pressure group, one that the ACLU has called "the most militant anti-gay organization in the northwest." In 1988, the group advocated a statewide vote to overturn an order protecting homosexuals from discrimination in state government.[9] The group has also threatened to run ultra-conservative third party candidates, forcing prominent Republicans to compromise on key issues of importance to far-right voters. Despite the collaboration of many conservative groups and political agendas, the *Washington Post* has noted that the OCA's organizing strength really comes from the group's "conservative Christian activists."[10] Under a banner of "traditional family values," the OCA continued such activism, including measures to force a vote on an initiative that would prohibit protection of homosexuals against discrimination and harassment. The OCA claimed 150,000 members in 1992.[11]

Even though the OCA was founded to advocate issues of importance to religious voters, the group's funding stream seems to have significantly broadened its ideological platform, according to a 1993 report by Dave Mazza. After years of subsidy by anti-environmental groups, the OCA drafted a statement of principles that not only reflected their New Right agenda, but it also, according to Mazza, included "several articles which dovetailed with the philosophical direction the nascent Wise Use Movement was going: privatization of government where possible, free market economy, nearly absolute private property rights and the conviction that the environment was primarily for the use of man." Mazza goes on to note that the "religious right, wise use, industry and other forces are tapping the same population pool, and that ties between these various factions are being created at the grassroots level."[12]

It is this kind of complex relationship that defines much of the Wise Use movement's current work. In exchange for funding, resources, or connections, the Wise Use movement has gained a large base of committed conservatives, even if such conservatives would not have initially identified themselves as anti-environment. However, having already made an impressive connection to fundamentalist Christian organizations, the Wise Use movement stands to gain considerably as the current political administration panders to Christian voters.

An instructive 1992 story in *Time* magazine highlighted the fact that the Wise Use movement always stands to gain in times of economic downturn. In the search for easy solutions to a troubled economy, many can be convinced that loosening some environmental regulations is a good thing. Indeed as *Time*'s Charles Alexander reported: "The Wise Use movement hopes to gain the upper hand by presenting itself as the voice of moderation in difficult economic times."[13] Given the importance the Bush administration places on the conservative Christian vote, and the continuing economic stagnation in the U.S. economy, Wise Use might easily grow in size and influence. The movement has, without a doubt, been reinvigorated and strengthened.

CHAPTER 2

WISE USE IN
THE DOMINION

In his book, *Crimes Against Nature*, Robert Kennedy, Jr., argues that the Wise Use movement became even more potent, both as an ideological and as a fundraising force, when it joined forces with the New Christian Right. In particular, he writes:

> ...when the Wise Use allies hooked up with Pat Robertson's Christian Coalition, they hit a home run. Robertson's special contribution to right-wing theology was to substitute environmentalists for communists as the new threat to democracy and Christianity.... They invoke Christianity to justify the rape of the land, violating manifold Christian precepts that requires us to be careful stewards. Rather than elevating the spirit, their interpretation of Scripture emphasizes the grimmest vision of human condition.

They embrace intolerance, selfishness, pride, arrogance toward creation, and irresponsibility to the community and future generations.[14]

This strategy combined fundamentalist Christians with some of the worst polluters in the land, and both assumed real power after they proved key to Bush's success in 2004. But while the organizational prowess of the New Christian Right has been generally acknowledged in helping deliver the 2004 election, little attention has been paid to its inherent fundraising abilities on display in that campaign, and its ability to thus influence policy.

In an interview with me, Kennedy asserted that Ron Arnold was key to making the interaction with Pat Robertson and other fundamentalists succeed. "Earth Day happened in 1970," Kennedy told me, "and soon after, environmental protection laws began to be put into place. Polluting industries saw that these laws would threaten their profits. Ron Arnold was instrumental in going to them and convincing them of a way to create an 'army' to combat the environmental movement." Arnold's strategies, said Kennedy, sparked a means of connecting fundamentalist Christians and the industries that stood to lose money because of environmental protections.

Kennedy, a practicing Catholic, also questioned whether the religious impetus in the policy-making of the Bush administration is genuine. The fundamentalism advocated by the administration, he said, "is not a religion. Religion is an organized framework for seeking truth. Fundamentalism is about power. Shakespeare says that the Devil quotes the Bible for his own purposes. With the fundamentalists, it's really all about power."

Mark Crispin Miller, however, disagrees, at least on whether the religious orientation is sincere. What's more, he says that the particular religion of the Bush administration makes its sincerity all the more fearsome. In an interview with Buzzflash.com, Miller, the author of *Cruel and Unusual: Bush/Cheney's New World Order*, made a direct connection between the administration and "End of Days" theology: "What's most significant here, and yet gets almost zero coverage in our media, is the fact that Bush is very closely tied to the Christian Reconstructionist movement," Miller told Buzzflash. "The links between this White House and that movement are many and tight. Marvin Olasky—a former Maoist who is now a Reconstructionist—coined the phrase 'compassionate conservatism,' and was hired by the Bush campaign in 2000 to serve as their top consultant on welfare...."

Miller defined the Bush administration fundamentalism as "Christian Reconstructionism," which he calls "a maverick theological movement."

"It's far more activist and radical than most Christian Evangelism is," Miller explained. "For the most part, Christian Evangelicals generally have chosen to deplore this world in their expectation of Jesus' return, whereupon this world will be improved. The Reconstructionists believe that it is the obligation of every Christian to do whatever he or she can do to make this a Christian republic with an eye toward making the other nations of the world Christian republics."[15]

Reconstructionism is the most common form of dominion theology, which is why both terms, in fact, are often used interchangeably to describe a fundamentalist Christian world view that advocates an activist stance based on a strict, literal interpretation of the Bible. Essentially, dominionists believe that the Bible is to be taken literally, and that the world is to be governed by what they call not a theocracy but a theonomy— that is, ruled not by God but by the law of God set forth in the Bible. Based on their reading of the Book of Revelations, they believe that once that rule is established around the world, and once Christianity has ruled the world for 1,000 years, Christ will return and all good Christians, living and dead, will ascend to heaven in what is called "The Rapture." (Some dominionists say Christ will return first, *then* there will be a 1,000 year "utopia" before the Rapture.)

The precise link between the environment and dominion theology lies in a popular interpretation of a well-known passage from the Bible's Book of Genesis: "God said unto them, Be fruitful, and multiply, and replenish the earth, and subdue it: and have dominion over the fish of the sea, and over the fowl of the air, and over every living thing that moveth upon the earth." (The passage was, according to the dominion theologists themselves, written by Moses,* along with the rest of the Book of Genesis, in 1445 BC.[16])

Even within the Conservative Christian movement, however, there are differences of opinion over what the passage means. Some say that it was a directive, in a time of wars and pestilence and climactic catastrophes, to build up decimated populations—a bit of encouragement to take heart and be strong. Others view it as a divine command

---

* Most other scholars date Genesis as having been written by authors unknown in about 460 BC. It's one of many differences: Many Conservative Christians maintain that the world is 10,000 years old, while scientists say that it's actually about 4 billion years old.

for stewardship of the earth. As a statement on the website of the Evangelical Environmental Network states, "Most major environmental problems such as air pollution, water pollution, and the threat of global warming hurt people. These problems fight against Christ's reconciliation of all of creation. In many instances they hit the poor, the children, and the elderly the hardest."[17] The organization, which publishes a newsletter called "Creation Care," includes numerous citations on its website of other sections of the Bible reinforcing the idea that Christians need to be environmentalists, too. Among the many is one, Jeremiah 4:18-28, picturing an angry God:

> '...your own conduct and actions have brought this upon you. This is your punishment. How bitter it is! How it pierces to the heart!'...Disaster follows disaster; the whole land lies in ruins...'My people are fools; they do not know me. They are senseless children; they have no understanding. They are skilled in doing evil; they know not how to do good.' I looked at the earth, and it was formless and empty; and at the heavens, and their light was gone. I looked at the mountains, and they were quaking; all the hills were swaying. I looked, and there were no people; every bird in the sky had flown away. I looked, and the fruitful land was a desert....[18]

However, there are those who take the Genesis passage to mean something entirely different: that Man has the right to rule over the natural world and use it (or use it up) as he sees fit. This view, when combined with the belief that the End Times are near, leads some to believe that either there is no need to take care of the environment, or, alternatively, that actively exhausting the environment will speed the Second Coming.

As a Counterpunch.org report by Joe Bageant puts it, "Reconstructionist doctrine calls for the scrapping of environmental protection of all kinds, because there will be no need for this planet earth once The Rapture occurs."[19] Or, as noted environmental reporter Glenn Scherer has observed, "Many Christian fundamentalists feel that concern for the future of our planet is irrelevant, because it *has* no future. They believe we are living in the End Time, when the son of God will return, the righteous will enter heaven, and sinners will be condemned to eternal hellfire. They may also believe, along with millions of other Christian fundamentalists, that environmental destruction is not only to be disregarded but actually welcomed—even hastened—as a sign of the coming Apocalypse."[20]

Although this may sound like the extremist views of a fringe group, the belief in the End of Times scenario spelled out in the Bible's Book of Revelations is actually widespread. In fact, a solid majority of Americans believe it, if recent polls are to be believed. A *Time Magazine*/CNN poll conducted in 2002 reported that over 59% of Americans take the prophecies of Revelations literally. A 2004 *Newsweek* poll put it at over 55%. For the majority of Americans, in other words, there is an obvious lack of incentive to preserve the environment, particularly amongst those who believe that Revelations' guaranteed ending will happen soon. Why concern yourself with fossil fuels or their emissions? Or destroying the greenbelt habitat, so crucial for climate and species protection? Why worry about what to do with hazardous waste, or nuclear waste? For that matter why even worry about the use of nuclear weapons?

It is a belief not only held by more than half the population, as the polls depict, but also by many of the fundamentalists leading the current administration or advising it on the shaping of environmental policy. (Glenn Scherer observes, "We are not talking about a handful of fringe lawmakers who hold or are beholden to these beliefs."[21]) James Watt, for example, remarked in his Senate confirmation hearing in 1981 that "I do not know how many future generations we can count on before the Lord returns," marking the self-professed born-again Christian as a dominionist. In fact, numerous statements from Watt made the point. Evangelist Austin Miles, in his book *Setting the Captives Free: Victims of the Church Tell Their Stories*, says Watt stated that, insofar as the environment is concerned, "God gave us these things to use. After the last tree is felled, Christ will come back."[22] Obviously, what this may mean isn't precisely clear. It could be read simply as an observation of reassurance—"God has given us the resources we need"—or something a bit more ominous and prompting: "Hurry up and exhaust the environment so Christ will come back." Either way, it points to the core belief of dominion theology: that Christ will return when our natural resources are used up. Put it in the mouth of a sitting cabinet secretary and it achieves an even greater level of resonance.

Which was no doubt why Bill Moyers cited Watt's statement in a Harvard speech about the endangered environment—or more precisely, he cited the fact that an article in *Grist*, an online magazine about the environment, had used the quote; *Grist* reporter Glenn Scherer subsequently cited Miles' book as the source. In any event, Moyers' December 2004 speech was widely reported on, reprinted, and circulated on the Internet. However, in an open letter, Watt denied having made the statement.* Responding to the Biblical interpretation attributed to Watt, the National Council of Churches (NCC) made a public statement on February 14, 2005 that called such Biblical thinking a "false gospel" and called upon Christians to reject its message. The NCC went on to say that "the sobering truth is that we hardly have knowledge of, much less control over, the deep and long-term consequences of our human impacts upon the Earth." However, the NCC warned that such an interpretation "still finds its proud preachers and continues to capture its adherents among emboldened political leaders and policymakers."[23]

Of course, despite the "proud preachers" of such a "false gospel" such as Watt, there are, as mentioned, those among the "End of Times"

---

* It was further pointed out that Moyers had prefaced the quote by saying Watt had made his statement before Congress, which was accurate to what Scherer had written in *Grist*, but not to what Miles said in his book. Scherer admitted his mistake in a correction, but it was Moyers who took the heat: He was roundly vilified as a "liar" across the right-wing media, even though he apologized to Watt, both in private and publicly, for parroting the error. It has not been disproved that Watt made the statement—it is simply his word against that of another evangelical, Austin Miles. However, given Watt's similar statements on the topic and his record on the environment, both as Interior Secretary (when he opened up millions of acres of public land to drilling and other resource extractions) and since (as one of the founders of the Mountain States Legal Foundation, a think tank dedicated to dismantling environmental protections, the statement seems to represent Watt's views in general. Watt's letter complained in particular that Moyer's use of the quote signified that the "religious left's political operatives have mounted a shrill attack on a significant portion of the Christian community."

believers who nonetheless interpret scripture to mean that we are "stewards" of God's creation and must protect it. One minister who preaches exactly that is Dr. John Shouse, Professor of Christian Theology at the Golden Gate Baptist Theology Seminary (a member of the Southern Baptist Conference) in Mill Valley, California. Shouse says that the Church could be doing more to encourage its congregations to be protecting the environment. The Bible is clear in its mandate, he told me in an interview, to "take care of God's work."

Not that he doesn't have his differences with secular environmentalists. According to Shouse, the typical environmentalist believes they are "one with the earth." However, Shouse told me, "The Bible clearly teaches that humanity is both a part of God's created order while also unique in it."

Such fine distinction are not made by many evangelicals, though. Many cite Genesis 13:17, often referred to as the "Doctrine of the Curse," to explain why there's no sense working with environmentalists, Christian or otherwise. "The Lord tells Adam, 'Cursed is the ground because of you.' That's the Doctrine of the Curse, and it means that God cursed all of creation," Calvin Beisner of Covenant College in Lookout Mountain, Georgia explained to Montana's *High Country News* reporter

Jeffrey Smith. "...so it's a fallacy to believe, as the evangelical environmentalists do, that nature untouched by the hand of man is pristine and pure. In fact, nature untouched by human hands is not as good as it can and should be."

According to Smith, Beisner "says that the Industrial Revolution 'liberated mankind from dependence on nature,' and that it is through economic progress that we redeem ourselves. As he explains how our 'cursed' creation will be redeemed, the talk turns from Scripture to the same sort of pro-business apologetic that evangelicals like [Pat] Robertson, [Ralph] Reed and James Watt advocate."[24]

Nor are the tenets of the Doctrine of the Curse the most extreme beliefs in the fundamentalist community. Reconstructionists led by R.J. (Rousas John) Rushdooney, one of the founders of the movement, hold that capital punishment should be meted out to homosexuals. Also, to women who have sex outside of marriage, women who have abortions and the people who advise them. Also, to astrologers. Their preferred method of execution is "stoning," for its Biblical accuracy and because it encourages community participation in the killing.[25] Rushdooney's son-in-law, Gary North, has written extensively about the tenets of

Reconstructionism. Or, as Declan McCullagh put it in a January 7, 1999 *Wired* magazine story, "For decades, Gary North has made a living predicting modern society will end in panic and ruin." As McCullagh points out, in 1980, North predicted "rationing of housing and a nuclear war with the Soviet Union." In 1987, as McCullagh quotes him, North said that AIDS would be the beginning of the end: "In 1992, we will run out of available hospital beds.... The world will eventually panic." In 1999, he predicted the Y2K computer glitch would lead to the collapse of the banking system and the aforementioned world-wide panic. While waiting for one of his predictions to come true, North regularly advises readers—often in his newsletter, *The Remnant Review*, where one subscription gets you 24 issues—to prepare for the end of the world by buying gold, grain, and guns and heading for the hills to await the return of Christ after the holocaust. Meanwhile, North has also written extensively about his belief that God has ordered all nations on earth to be under Christian dominion, led by a theocracy here in the United States that will rule until the "End of Times." To North, America's wars around the globe are in perfect synch with Biblical prophecy, proving that we are now in the beginning stages of this "End of Times."

The complexities of domnionist/reconstruction-ist Biblical interpretation—and the subtleties of theonomy—seem to require book after book of such explication. Another leader of the reconstructionist movement Kenneth Gentry Jr., an Orthodox Presbyterian minister who is chancellor of Christ College in Lynchburg, Virginia, is an immensely prolific author. Among his many titles on the tenets of Reconstructionism are, to name just a few, *Yea, Hath God Said: The Framework Hypothesis / Six Day Creation Debate; The Christian Case Against Abortion; God's Law in the Modern World; The Beast of Revelation; Before Jerusalem Fell: Dating the Book of Revelation; He Shall Have Dominion: A Postmillennial Eschatology;* and *The Greatness of the Great Commission: The Christian Enterprise in a Fallen World.*

But the rigid system of the End Times envisioned in the writings of Gentry, Rushdooney, and North has led several critics, most notably religion expert and former Catholic nun Karen Armstrong, to cite the potential for fascism in the Reconstructionist movement. In her *The Battle for God* she cites Rushdooney and North in particular as espousing a worldview that "is totalitarian." "There is no room for any other view or policy, no democratic tolerance for rival parties, no individual freedom," she writes.[26]

Chip Berlet and Matthew Lyons, in their book *Right-Wing Populism in America*, are blunter still: They refer to the movement as a "new form of clerical fascist politics."[27]

While it is difficult to determine to what extent dominionist thinking actively concerns the environment, a few central points are clear. For one, the Wise Use movement has actively courted, and organized alongside, Christian fundamentalists of all varieties. Indeed, it is increasingly difficult to determine why certain religious groups oppose environmental protections. It seems certain that some fundamentalists oppose these protections because of a dominionist understanding of the "End Time." As noted, the National Council of Churches, for one—an organization representing 45 million people and 100,000 local congregations—judged the belief widespread enough to respond to, and refute, its teachings in detail.[28]

Perhaps the real point to be made, however, is that America today remains a breeding ground for extremist versions of Christian fundamentalism. Even for many devout Christians, widespread Fundamentalist teachings about the Bible seem extreme. It is a problem, especially as reconstructionist and dominionist ideas continue to be promulgated, that will only get worse with time. Indeed, as Glenn Scherer has pointed out, Christian children are often reared on Reconstructionist textbooks, among them *America's Providential History*, which teaches that:

> The Christian knows that the potential in God is unlimited and that there is no shortage of resources in God's Earth. The resources are waiting to be tapped.... The secular or socialist has a limited resource mentality and views the world as a pie...that needs to be cut up so everyone can get a piece.[29]

This belief alone—which is being fed to an unknown number of schoolchildren—teaches that the world's resources are sufficient, that there is no need to protect or fret about the environment. The dominionist paradigm talks of "limited resource mentality," as if environmentalists lacked a proper imagination. As the textbook explains further: "While many secularists view the world as overpopulated, Christians know that God has made the earth sufficiently large with plenty of resources to accommodate all of the people."

CHAPTER 3

THE "GRASSROOTS" CAMPAIGN AGAINST THE GRASSROOTS

The most prominent—certainly the biggest, and probably the most powerful—"citizens" organization in the Wise Use movement is the Alliance for America (AFA), a coalition that claims a membership of over 500 groups spread out over the entire fifty states. Founded in 1991 with funding by the Cattlemen's Association, the American Mining Congress, the Petroleum Institute, the Chemical Manufacturers Association, and other industry groups, as well as with money from the Reverend Sun Myung Moon-funded American Freedom Coalition, the AFA is headed by Rose Comstock Correa, the former owner of a timber mill who first came to prominence fighting protections for the endangered Spotted Owls of the great Northwest.

Comstock's biographical note on the AFA website highlights a stance common among Wise Use advocates: the belief in the hyper-protection of private property alongside the denial of protection for public property—an attitude of what's mine is mine, and what's yours is mine, too. As Comstock's bio puts it, she is a "long time activist supporting multiple use of public lands and private property rights protection."[30]

The AFA membership, meanwhile, highlights the intersection between Wise Users, fundamentalist Christians, and the far-right, with members including groups dedicated to eliminating endangered species protections such as the American Land Rights Association; others who want to get rid of tree cutting restrictions such as Potlatch; the Blue Ribbon Coalition of off road vehicle users; groups that want to open up the Alaska National Wildlife Arctic Reserve (known as ANWAR) to oil drilling, such as—and this is no typo—ANWAR.org, yes, the Alaska National Wildlife Arctic Reserve; and broadening out, groups that don't believe in government or taxes, such as the Property Rights Foundation of America or the Heartland Institute.[31]

The AFA, whose slogan is "We are the *true* conservationists," sponsors what it calls "Fly-In for Freedom" events in Washington, DC, where members are matched up with government officials who are expected to hear their case. In 1992, the group sponsored a protest and 12-hour vigil outside CBS News's Washington bureau, claiming that the network had produced " 'flagrantly biased' reporting against loggers, ranchers, miners and private property rights." (Amongst their protest signs at the vigil: "Put people first," and "Honk if you use toilet paper.")[32]

The group does its best to appear small; members of the AFA claimed in 1993 that the organization had only $20,000 in the bank.[33] But from the very beginning, the group has had close ties to industry cash. The *New York Times* reported in 1991 that the group was associated with Alan M. Gottlieb, a conservative fundraiser who at the time ran a vast direct-mail network.[34] But it is also true that part of the genius of the AFA is that it doesn't need much money, for it is, in reality, simply another layer in an onion of Wise Use groups, many of which industry supports directly, and most of which lead back to a handful of Wise Use ideologues. The AFA usually maintains that participants of the "Fly-In for Freedom" tours pay their own travel expenses. In reality, most were sent by member organizations. The AFA offices, on the other hand, have "sophisticated computer technology... that allows regular updates to members on issues before Congress."[35] Gottlieb claimed in 1991 that he found it very easy to raise money for AFA's members—he noted that "the environmental movement has become the perfect boogyman."[36] Gottlieb is one of Ron Arnold's closest associates and the founder and president of the Center for the Defense of Free Enterprise, where Arnold is executive vice president. In 1995, the AFA dropped its "poor-house" ruse and hired Edelman Public Relations to push its legislative agenda to the media and other parties.[37]

The organization has grown steadily since its founding in 1991, when it had 125 member organizations. By 1992, the AFA had 400 member organizations; that number may be as high as 650 today, connecting the AFA to a reported 5 million people.[38] The organization's impact has grown as well. Jerry Schill, president of the North Carolina Fisheries Association and an AFA officer, notes that the organization has also seen a steady increase in power, saying "What started out as a gnat that could be easily swatted away has grown into a mosquito that's sucking blood."[39] In 1991, the AFA succeeded in getting Congress to pass a bill to allocate a portion of the Federal gas tax to build trails on Federal lands for off-road vehicles (such as the project in the Stanislaus National Forest that Judith Spenser was fighting). Under Clinton, the group claimed responsibility for killing a broad-based energy tax, one that the AFA believed would cause significant job losses.

However, unless the AFA declares victory on a particular piece of legislation, it is difficult to gauge the group's real impact. In the case of the failed energy tax, for example, the work of the AFA and other Wise Use groups was aided by direct lobbying from a coalition of energy and manufacturing interests. Indeed, it seems that while groups like the AFA can often claim an ideological victory, it is unaffiliated private interests that immediately benefit from such activism.

The Wise Use movement's emphasis on private use of public resources means that corporate interests stand much to gain from the ongoing advocacy work of Wise Use organizations. Among these interests, resort and theme park developers, off-road vehicle manufacturers, timber industries, and a wide array of land developers stand to gain in particular. In fact, these are the very interests that financially support both Wise Use and right-wing Republican politicians and the Bush administration in particular.*

Charles "Chuck" Cobb, Jr. is an example of someone who bridges many of these areas. Former Under Secretary of the U.S. Department of Commerce under President Ronald Reagan, appointed Ambassador to Iceland by President George H.W. Bush, Cobb currently serves as chair of Florida FTAA, Inc. (the Florida branch of the Free Trade Area of the Americas organization), an appointment given him by Florida Governor Jeb Bush. Cobb, together with his wife, Sue, who was appointed Ambassador to Jamaica in 2001 by George W. Bush, have long been major donors and fundraisers for the Bush family.* In private life, Cobb was for many years Chairman and Chief Executive Officer of Disney Development Company and served as a member of The Walt Disney Company's Board of Directors and on its Executive Committee.

Now, Cobb is part-owner of the Kirkwood Ski Resort, which is located on public land in the Sierra Nevada Mountains in California (another

* For example, Exxon/Mobile, one of the leading corporate contributors to the Bush presidential campaign in 2000, and again in 2004, also gave Ron Arnold's think-tank, the Center for the Defense of Free Enterprise, $130,000 in 2003 for work on Global Climate Change issues.
* According to People for the American Way, the Cobbs gave $300,000 to the Bush presidential campaign in 2000.

co-owner, until 2004, was former Vice President Dan Quayle). Cobb and Kirkwood won approval to develop another project on public land, this time on Caples Mountain in the El Dorado National Forest, also in the Sierra Nevadas. In a deal that was not publicly announced until after it was approved, that pays virtually nothing to the Federal government, and waives the payment of all local and state taxes, Cobb received clearance to build a panoramic restaurant and resort atop what had been the pristine wilderness of Caple Mountain.[40] As the developer's website describes it, "The introduction of a mountain-top restaurant at Caples Crest will provide breathtaking views of Caples Lake, exquisite dining and year-round recreational activities via its high-speed lift connection to the village."[41]

Under close scrutiny, Cobbs' ties to Disney are by no means inconsequential here. Disney has long had its eye on public land. In the late Sixties, for example, the U.S. Department of Forestry approved a plan that would have allowed Disney to build a theme park in the Sequoia National Forest. The Sierra Club sued, and eventually managed to get Sequoia out from under the jurisdiction of the Forest Service and placed under the more protective status of the National Park Service, thus narrowly blocking the effort. A description in the court judgment handed down in 1972 detailed what Disney had in mind: "The final Disney plan, approved by the Forest Service in January 1969, outlines a $35 million complex of motels, restaurants, swimming pools, parking lots, and other structures designed to accommodate 14,000 visitors daily. This complex is to be constructed on 80 acres of the valley floor under a 30-year use permit from the Forest Service."[42]

More recently, in 1995, the Disney Company signed a "Memorandum of Understanding" with the Forest Service. (Also in on the agreement was the Department of Agriculture's Natural Resources Conservation Service, the Army Corp of Engineers, the Department of the Interior's Bureau of Reclamation, the Fish and Wildlife Service, and the National Park Service.) A similar agreement was struck in 2003. One of the express purposes of these documents, as the 1995 agreement explained, is "to work together in partnership on issues of common interest and upon which the cooperators can jointly plan and carry out mutually beneficial programs and activities consistent with each organizations mission and objectives."[43] And in September, 2004, Interior Secretary Gale Norton awarded Disney, along with the Toyota company, a "Take Pride in America" award for being among "those who best protect and/or enhance our public parks, forests, grasslands, reservoirs, wildlife refuges, cultural and historic sites, local playgrounds, and other recreation areas." [44]

All of which leads many environmentalists to surmise that it's only a matter of time before Disney attempts another takeover of a national forest. Indeed, Arnold's *Wise Use Agenda* called for handing over the National Park system to private enterprises "with expertise in people-moving such as Walt Disney." And Secretary of the Interior Norton has also called for such "outsourcing" of National Park Service jobs to companies such as Disney which are geared to provide "better delivery of services to the public."[45]

Meanwhile, with former Disney executive and board member Chuck Cobb now given official sanction to develop the El Dorado National Forest, when his wife held a position in the administration of the president and he himself holds a position in the administration of the president's brother, those environmentalists may feel haunted by a Disney song that's particularly apt—the one about it being a small world after all.

Books have proven a powerful medium to transmit the message of the dominionist movement, so much so that on his website, Kenneth Gentry offers a correspondence course called "Righteous Writing," which he says "provides instruction on reading techniques, research method, improving writing style, securing a publisher, and techniques of book promotion."[46]

Perhaps the most successful books in recent American history, indeed, some of the most successful books in all of American history, have in fact been books whose purpose is to proselytize the reconstructionist "End of Times" message—that is, the *Left Behind* series written by Tim LaHaye and Jerry Jenkins, which, according to a February 2005 *USA Today* report by Carol Memmot marking the series' 10th anniversary, has sold over 70 million books. Though presented as fiction, the books faithfully represent the reconstructionist's "End of Times" scenario, in which all the predictions of the Book of Revelations come true—non-believers (such as Jews and Muslims) are "left behind" as the reconstructionists and dominionists are swept up to heaven in the Rapture. For an additional $29.95, fans can join LaHaye's "Interpreting the Signs" website, where they can interact with LaHaye and other Reconstructionist Christian evangelists who will help them prepare for Armageddon.

Meanwhile, according to a January 2004 report by Robert Dreyfus that appeared in *Rolling Stone*, LaHaye is "a man who played a quiet but pivotal role in putting George W. Bush in the White House."[47] For example, often overlooked in newspaper features about LaHaye is that, in 1979 along with the Rev. Jerry Falwell, he co-founded the Moral Majority—the group that acted as the organizational infrastructure for bringing the right-wing political world into alliance with the Christian Right. (And bringing it into alliance, as well, with some big money men from the Christian Right, including one of the biggest funders of the Moral Majority, beer magnate Joseph Coors, as well as other Moral Majority donors including Amway founder Richard DeVos, and Texas oil men Nelson Bunker Hunt, a John Birch Society member and founder of the Christian World Liberation Front, and T. Cullen Davis, who gained national notoriety when he was charged with brutally murdering his ex-wife's boyfriend and daughter—he was found innocent, despite seemingly strong evidence, in a sensational murder trial.) In 1981, LaHaye co-founded, and served as first president of, the Council for National Policy, a group ostensibly set up to be a conservative alternative to the Council on Foreign Relations, although the CNP does share members with the CFP; George Gilder, Edward Teller, and Guy Vander Jagt, all CFR members, were on the original CNP board. The group regularly advises the White House on a wide range of issues and tactics (Robert Dreyfus contends that the plot to impeach Bill Clinton was hatched by the CNP at a June 1997 meeting in Montreal) and includes among its members Pat Robertson, Oliver North, Ralph Reed, Phyllis Schafely, Pat Boone, and Nelson Bunker Hunt.[48]

LaHaye has various other little-noted ties that place him at the very heart of the New Christian Right movement, starting with his graduation from the unaccredited Bob Jones University in South Carolina, the controversial fundamentalist Christian school that came into the spotlight when George W. Bush spoke there during his first presidential campaign, despite a public outcry that emanated even from many conservatives.*

---

* The controversy centered on the school's decades-long banning of African Americans from enrollment until forced to do so by a 1971 court order, and for its anti-Catholic teachings—founder Bob Jones considered Catholicism a "satanic cult." Bush's February 2, 2000 appearance there no doubt helped him win a hotly-contested primary in the conservative state, but the national outcry over his appearance continued until the campaign issued a statement specifically denying that Bush was a racist or an anti-Catholic (thereby seeming to confirm liberal charges that BJU was indeed a racist institution). On February 26 Bush also wrote a letter of apology to Cardinal John O'Connor.

But whatever impact LaHaye has had on the attitude and policy-making in the White House pales in comparison to the grass-roots impact he has had in spreading the dominionist / reconstructionist philosophy of Armageddon via the *Left Behind* series. The phenomenal success of the series is perhaps the single-most powerful factor behind the rapidly growing belief in an "End of Times" theology amongst average Americans, which in turn has no doubt contributed immeasurably to a drastic lessening of concern about the environment.

That an "End of Times" fundamentalism is entwined with Republican politics on a pervasive, grassroots level is now undeniable. I bore firsthand witness to it in my own community in supposedly liberal Northern California when I attended a luncheon of the Federated Republican Women of Marin County featuring a talk by Dr. William Wagner. Wagner is a long time fundamentalist missionary, a reconstructionist evangelical who has worked, he says, all over Europe, Africa, and the Middle East. He is also the author of *How Islam Plans to Change the World*, a book that details his conception of a Muslim plot to take over the planet, and is currently under wide discussion on right-wing radio. (Wagner's book is published by Kregel Publications, who also publishes Kenneth Gentry, Jr.) Such was the topic of his talk to the Republican ladies, delivered as a passionate sermon full of the fore-boding, ulterior motives of the Muslim community, both at home and abroad. With no trace of irony, he described the Muslim plot to do, in other words, exactly what the reconstructionists such as Rushdooney, North, and Gentry advocate in their writings: have their strict, fundamentalist religious doctrine take over the world. Wagner, who called himself a "card-carrying Republican" and boasted of a long-time friendship with Republican Senator Pete Domeninci, also said he believed there were actually four "religious groups" with take-over-the-world "mega-strategies": The four include Muslims, Mormons, his own Southern Baptists, and homosexuals. (He did not explain how homosexuality constitutes a religion.)

It was, in the end, not only an impressive promulgation of the "End of Times" scenario as a necessary one to fight off an evil enemy, but also an effective outreach of support for the administration's ongoing wars in the Middle East—the modern-day mix of ideology and politics in powerful microcosm. It led more than one of the Republican women to become somewhat agitated. One bespectacled, gray-haired woman exclaimed, "Dr. Wagner! Dr. Wagner! You've been describing The End of

Times! Aren't you describing the End of Times?" All eyes landed on Wagner, awaiting his response. It seemed as if he was calculating his answer carefully. Finally, he said, "Why yes, yes, the Prophecies do talk about this." The woman, on her feet now, persisted, "Do you think God put Bush in the presidency to lead us?" "Why yes," Wagner replied, "it is most certain God put Bush in office." The women in the room nodded their agreement, and started conversations amongst each other. They were shushed by one of their leaders so that Wagner could go on. He launched a depiction of American Indian reservations as "terrorist cells," and described terms such as "multi-culturalism" as code words for terrorists.

The woman who had become so agitated consented to an interview with me afterwards in her home in a comfortable suburb of San Francisco. As it turned out, the woman—I'll call her by the pseudonym Margaret, out of deference to the probability that her fellow Republican Women would be critical of her for speaking with me—was a retired schoolteacher, widowed with four grown children, and a passionate fan of the entire oeuvre of Tim LaHaye. A former Catholic, Margaret now calls herself an "Orthodox Presbyterian." Also known as "Reformed Presbyterians," this branch of the

Presbyterian church shares with reconstructionism a belief in the literal word of the Bible.

Besides LaHaye, one of the main inspirations guiding Margaret's belief that we are in the "End of Times" was the writings of John J. Alquist, founder of the website The First Internet Christian Church. Margaret first encountered Alquist's writings in a newspaper ad for himself.[49] Alquist, also a former Catholic, is a St. Petersburg, Florida Pentacostalist who also sells and distributes Amway-like products (motor oil, fruit juices, portable infrared saunas) through his First Internet Church website. On the part of the site constituting his "cyber-church," Alquist details the signals that we are in the End of Times via some of the stereotypical reasons—the existence of the state of Israel, a drop in morality, various wars and natural disasters. There is a passage citing the "mark of the beast," which made me wonder if Margaret was aware that many reconstructionists consider the omnipresent bar codes of daily commerce this very "sign." And to the list of usual suspects that the site lists, Alquist, with an all-too-familiar lack of irony, adds a new sign of his own devising: the rise of the Internet.

Alquist also cites Timothy 4:1 to say that "God is not part of nature, the core belief of pantheism"—

which reminded me of Dr. John Shoue's admonition that "humanity is both a part of God's created order while also unique in it," i.e., that we are not "one with the earth" as environmentalists believe.[50]

Margaret told me she agreed with this, and that she equated the environmentalists' attitude with a worship of nature, which she considered idolatry. She listed Egyptians and Native Americans among the people who are disobeying the Bible in this way. "Many people worship the way Indians used to worship and that is wrong," she told me. "They worship Nature, and that's just wrong. This New Age religion is very dangerous, and they teach people to worship things the Egyptians worshiped, and that's against Jesus Christ."

Still, she told me, her belief that we are in the "End of Times" did not mean we shouldn't protect the environment. She believed it was her Christian responsibility to take care of the Earth, and she cited several passages from the Bible backing this up: "And God made the plants and the animals and saw that it was good. That's in Genesis."

Margaret was, in fact, emphatic on the point. "God made the Earth and all the beautiful flowers and animals and he expects us to take care of it for him for when Jesus returns," she said. "He's not going to like it if we destroy what he made."

Furthermore, as a Republican, Margaret told me that "I feel strongly [about preserving the earth] as a private property owner, too. Why, when my children inherit my land, I want it to be beautiful for them and for my grand children. I think everyone wants that."

When I pressed her by citing numerous pieces of environmentally harmful legislation cited by prominent Republicans, many of them self-professed Christians, she thought for a moment, then responded, "They aren't really Christians."

And the President? I listed some of his policy decisions, appointments, and statements.

This led to a longer pause.

Then Margaret leaned forward and looked me deeply in the eyes.

"George W. Bush is a *good* man," she said firmly. "He may get bad advice from people, but he is a good Christian, and I believe God is leading him in leading our country."

In the end, we came back to Tim LaHaye. Margaret brought out her copy of his *Prophecy Study Bible*, a 1,600 page behemoth of a book with easy-to-read color inserts that spell out, in great detail, the path to Armageddon.

According to LaHaye's reading of the Bible, the End of Times process is a complicated one, involving an Antichrist who will set up a "one world government" (apparently, not to be confused with the theocratic government reconstructionists say *should* rule the world), false prophets, plagues of boils (which many reconstructionists interpret to be AIDS), and the Euphrates drying up (which, in a way, did happen in the 1980s, when Saddam Hussein, with American assistance in what has been called a "major ecological disaster," dammed the river up in an attempt to dry up a vast marsh-land relied upon by a population that opposed his policies.[51] Post-Saddam Hussein, the region continues to undergo severe ecological damage thanks to the war, particularly from the American forces' use of depleted uranium and from the massive pollution of the numerous U.S. aircraft carriers now in the vicinity.)

LaHaye also describes the fabled "Rapture": All non-believers in Christ, such as the Jews, will be left behind while all believers, living and dead, will literally float up to a heaven in the sky. (The recon-stituted dead will go first.) And LaHaye is adamant: "there can be no valid system of biblical prophecy without belief in the Rapture."[52]

CHAPTER 4

WHO'S WISE IN GOVERNMENT?

The concept of Wise Use has resonated profoundly with the principles of, and become a tool for, the New Christian Right and its dominant dominionists. As a result, the Wise Use movement has spread beyond being a euphemism for the commercialization of public property.

As Montana journalist (and sometime *New York Times* and *Washington Post* correspondent) Mark Matthews observed in the *High Country News* as far back as February 1996, "For many environmentalists, the Christian Coalition is synonymous with the wise-use movement, a philosophy which adherents claim is supported by the Bible when it tells humans to 'subdue' the earth and take 'dominion over the fish of the sea, and over the fowl of the air, and over every living thing that moves upon the earth.' The notion that development of the land equals progress has been a mainstay in the nation's religious philosophy since Colonial times, when the Puritans saw themselves as creating a paradise in the wilderness. The Montana spokeswoman for the Christian Coalition, Laurie Koutnik, says her organization believes that God 'gave us the animals to use wisely.' The Endangered Species Act 'was well intentioned in the beginning,' she says, 'but now it is absurd. We need a balance with nature, but not at the expense of people's livelihoods and private property rights.' "[53]

Another early glimpse at this intersection was documented by Jeremy Leggett in his book *Carbon Wars*, a book about the fossil fuel industry's intersection with political policy. Leggett describes a 1997 talk with Ford Motor Company executive John Schiller in Geneva, Switzerland at a meeting for negotiating climate issues that was well attended by what Leggett refers to as the "Carbon Lobby"—people from the coal, oil, and other fossil fuel industries. Schiller, an emissions planning associate with the Emissions and Fuel Economy Office of the Environmental and Safety Engineering staff of Ford Motor Company, told Leggett, "You know, the more I look, the more it is just as it says in the Bible." Schiller went on to say that because of the Biblical prophesies, "it really didn't matter about climate change, anyway." He cited, in particular, the Book of Daniel and its prophecy of a New World Order being led by the Antichrist. When Leggett pointed out that the World Council of Churches had gone on the record as being concerned about climate change, Schiller replied, "I don't regard them as a Christian organization." Leggett concluded from the exchange that to Schiller, "Environmentalists, it seemed, were in league with the antichrist, whether wittingly or not."[54]

But is our fundamentalist government really composed of that many dominionists, or that many Wise Use adherents?

It would be best, of course, to address these issues directly with the dominion theologists themselves. But as Bruce Barron, an evangelical Christian and the author of *Heaven On Earth? The Social and Political Agendas of Dominion Theology*, told me in an interview, dominion theologists currently involved in making environmental policy are likely to be too stealthy for that. "You have to look at their actions, what they are doing, not necessarily what they say," he advised me.

Barron knows whereof he speaks—a former member of Pennsylvania Senator Rick Santorum's staff, he shares many of the beliefs of the Religious Right. He home-schools his children, believes homosexuality is a sin, and doesn't see anything particularly wrong with a blur of the lines between church and state. He is not himself a dominionist, however, and says that as far as the environment is concerned, he considers himself a "preservation-ist," and believes that because "God made nature" we are obligated to take care of it.

Nonetheless his own responses reflect some of the stealthiness he warns about: When I pressed him on whether or not he felt there was an apoca-lyptic ideology driving the environmental poli-cies of many of the Republican office holders, he pressed me in return: "Have they publicly stated that they believe that the earth has to burn for Christ to return?" For Republican office holders, he said, "I think it is really their world view that is influenced by the fact that they are politicians, always running for office." And for those Republicans, he said, "The environmentalists went too far. Their desire to protect the earth intruded on private property rights and free enterprise. You know, we differentiate between preservationists and environmentalists. I think that may of us feel that 'preservationists' are more reasonable than environmentalists, they're more willing to compromise."

To illustrate his point, he began a story about a "preservationist" advising Santorum that a hazardous waste incinerator being built near East Liverpool, Ohio would not be harmful to the surrounding community. But as it turned out, I had been the producer on that very story for a radio news show—a story on the opposition to the incinerator before it was built, and then, after-wards, on the unusually high rates (40% higher than the national average) of a rare form of eye cancer that had been discovered in children attending the school building next to the incinera-tor. Barron thought about it for a moment when I filled him in. "Well," he responded, "sometimes we don't know these things until after they go into operation, and that's how we find out the risks."

Still, there was something refreshingly honest in his core assessment of the motivating factors for the politicians of the religious right: "Unfortunately, I'm afraid with some of them, that Big Business has more of an influence than the Bible in many of the decisions they make. They've strayed too far from the true meaning of the Scripture to be 'Stewards of the Earth.'"

And of course, his suggestion that the actions of these religious men and women speak louder than words is the best way to cut through their obfuscation. The clearest answer to the question of how religious politicians interpret Biblical passages on concepts of dominion is in the policies they make and support. Are they making policy that follows Biblical precepts saying man is the guardian of the earth, or are they making policy that supports the concept that we are in an "End of Times" scenario?

In May of 2001, *Sierra Magazine* conducted exactly the kind of study Barron suggests, with a penetrating examination of how, after taking office in 2000, George W. Bush went about reorganizing the Department of the Interior (which oversees both the National Parks Service and the Bureau of Indian Affairs). "Of the 58 people appointed by President Bush to help shape the Interior Department," the magazine reported, "more than half are [timber and other] industry lobbyists, executives, and consultants, 12 are active in the anti-environmental 'wise-use' movement. The members of this 'transition team' make policy recommendations and assist in filling positions at the agency; some may be hired for staff jobs themselves."[55]

Prominent wise-users on that team included:

TERRY ANDERSON, executive director, Political Economy Research Center, a free-market think tank that aggressively advocates privatizing federal lands. It is listed as a networking participant on the Alliance for America website.

DEMAR DAHL, rancher and president of the Jarbidge Shovel Brigade, which received national attention in July of 2000 for illegally opening up a closed Forest Service road.

W. HENSON MOORE, president of the American Forest and Paper Association, a trade group for the wood products industry, which is also associated with the Wise Use umbrella organization, Alliance for America. Moore called Clinton's protection of 58 million roadless acres of national forest "a policy that confounds both science and common sense."

HAROLD P. "HAL" QUINN JR., senior vice president, National Mining Association. The NMA, like the groups led by Moore and Anderson, is part of the Alliance for America.

DIEMER TRUE, vice chair of the Independent Petroleum Association of America. True is also a board member of the Mountain States Legal Foundation, an anti-environment think tank which represents loggers, miners, ranchers, and developers fighting environmental safeguards.

Republican Senator TED STEVENS from Alaska, who has long advocated opening up more of the Alaskan wilderness to oil drilling. Stevens is also chair of the Senate's Commerce Committee. Although the Committee oversees the Senate's monitoring of global warming issues, Stevens disagreed with a recent scientific report commissioned by the committee. As the Associated Press reported on November 19, 2004, Stevens "said he does not accept the conclusion the scientists reached: that the driving force behind warming is people burning coal, oil and natural gas, the fuels that produce greenhouse gases that trap heat in the atmosphere." Stevens also receives significant support from the Christian Coalition.

Republican Congressman RICHARD POMBO from California, the co-chair of the House Resources Committee, has a long record of voting against environmental protective legislation and of working hard to weaken the Endangered Species Protection Act. [56]

People subsequently appointed to positions of power, and appointed to key advisorships, in the development of the Bush environmental policy were no less encumbered with connections to the Wise Use movement and the New Christian Right.

In a another investigation for *Sierra Magazine*, David Helvarg reported in September 2004 that the "veterans of the once-discredited militant anti-environmental groups are now setting natural-resource policy for the Bush administration."[57] He detailed the membership of both Interior Secretary Gail Norton and former Department of Agriculture Secretary Ann Veneman in the Wise Use think tank co-founded by James Watt, the aforementioned Mountain States Legal Foundation. Norton was hired as a lawyer for the foundation by Watt himself in 1979, and he has been called her mentor. Meanwhile, the Mountain States Legal Foundation has become a key center for developing litigation against environmental protections, including the dismantling of the Endangered Species Protection Act, advocating increased road-building in protected wilderness areas, and increased drilling for oil and mining for minerals on Federal lands. The group also lobbies Congress, funds Republican campaigns, and advises on the drafting of policy for the Bush administration.

Interior Secretary Norton has numerous other ties to what many would perceive as anti-environmental organizations, such as her former position as a lawyer with Brownstein, Hyatt & Farber, P.C., where she represented Delta Petroleum and lobbied on behalf of NL Industries, a paint manufacturer defending itself in liability suits over children's exposure to lead-based paint.[58]* On behalf of its clients, the firm also gave tens of thousands of dollars to mostly Republican campaigns during the 2004 election, and to members of Congress supporting anti-environmental legislation, including Senators Rick Santorium, Arlen Specter, Tom DeLay, Dennis Hastert, and others. Norton was also the founding Chair of the Coalition of Republican Environmental Advocates, a group funded by, among others, Ford Motor Company, oil giant BP Amoco (now BP America), National

* With the appointment of Stephen Johnson to head the Environmental Protection Agency, we see another example of this pro-business, anti-child policy, which proves to be one of the most glaring failures of Bush's second term. As Acting Administrator for the EPA, Johnson saw nothing wrong with his program "CHEERS," (Children's Environmental Exposure Research Study), a program with an offer like a game show: low income families would receive $970.00, a camcorder, and clothes, to spray pesticides in the primary room of their young children, ages three and under, for over a period of two years. After strenuous objection to CHEERS from Senator Barbara Boxer and Senator Bill Nelson, saying they would block his nomination, he was forced to cancel the program. Yet in his begrudging retraction statement, Johnson made no mention of the ethical or moral problems with pesticide experimentation on small children: "The Children's Health Environmental Exposure Research Study was designed to fill critical data gaps in our understanding of how children may be exposed to pesticides (such as bug spray) and chemicals currently used in households....the study cannot go forward....EPA must conduct quality, credible research in an atmosphere absent of gross misrepresentation and controversy." Johnson, a graduate of Taylor University, "an intentional, Christ-centered, learning, living and serving community," is quoted in his Taylor alumni profile saying "....what I appreciate now is the spiritual and moral grounding that Taylor provided."

Mining Association, the American Forest Paper Association, and the Chlorine Chemical Council, and she has served on the board of the National Policy Forum.

Norton is a member of the Federalist Society, an organization with significant ties to both the Wise Use movement and the New Christian Right. People for the American Way, among others, has cited the Society as being especially influential on President Bush's environmental policies.[59] For example, when Bush broke his campaign promise to cut carbon monoxide emissions, and subsequently pulled the U.S. out of the Kyoto accord on climate change, he did so after reading what he called "important new information." That information was a report commissioned by David McIntosh, a Federalist Society founder and former Representative from Indiana, arguing that toxic emissions were exaggerated and warning of costs to business from regulating those emissions. Donald Paul Hodel, former president of the Christian Coalition, is one of the many Federalist Society members who hail from the Christian Right. (A partial list of other members of the Bush administration who are also in the Society includes: former Attorney General John Ashcroft; Secretary of the Department of Energy Spencer

Abraham; Solicitor of Labor Eugene Scalia, the son of Supreme Court Justice Antonin Scalia; General Counsel of the Department of Education Brian Jones; Deputy Attorney General Larry Thompson; Solicitor General Ted Olson; Inspector General of Department of Defense Joseph E. Schmitz; and Assistant Attorney General for Environment and Natural Resources Thomas L. Sansonetti. Also, John Roberts, under nomination for Chief Justice of the Supreme Court as this book goes to press.)*

At the Department of Agriculture, Mike Johanns, the man who replaced Ann Veneman as Secretary, has an even more activist record in both Wise Use advocacy and in her own Christian fundamentalism. Johanns, as governor of Nebraska, was a strong proponent of industrial agriculture, which promotes heavy herbicide and pesticide use, often resulting in contaminated soil and water supplies, and both as governor and since becoming Secretary he has worked to limit federal oversight on environmental issues, leaving

* Other members, according to People for the American Way: Supreme Court Justice Antonin Scalia; Senator Orrin Hatch; former Independent Counsel Kenneth Starr, whose investigation led to the impeachment of President Bill Clinton; Robert Bork, the conservative judge whose nomination to the Supreme Court was famously blocked in a bitter and divisive Senate hearing; Linda Chavez, the FOX News commentator and President of the Center for Equal Opportunity who withdrew her nomination to be George Bush's Secretary of Labor; Michigan Governor John Engler; and Charles Murray, the "controversial author who asserted that some races are inherently less intelligent than others."

national public lands vulnerable to state control—often dictated by corporate or corrupted state government interests.[60] Meanwhile, as *The Washington Post* reported, "Johanns has come under criticism from civil liberties groups...for official actions that they said promoted conservative Christian beliefs. In May 1999, he signed a proclamation declaring a 'March for Jesus Day,' and he later endorsed a 'Back to the Bible Day' in honor of a fundamentalist Christian group in Nebraska. But he refused to sign a proclamation honoring 'Earth Religion Awareness Day,' an event organized by the Wiccan church. The church practices witchcraft as part of a pagan religion that U.S. federal courts have said is protected under the First Amendment. In explaining his refusal to declare the Earth Religion day, Johanns told reporters he would not sign 'something that I personally disagree with.' The American Civil Liberties Union criticized Johanns's position, and an official of the group Americans United for Separation of Church and State, said that the governor was 'picking one religion over another as a government official.' "[61]

Another close advisor within the White House is James Connaughton, a former oil and chemical industry lawyer, who is now one of the President's senior advisors on the environment as chair of the White House Council on Environmental Quality. As the Center for Media and Democracy notes, "His office pressured the Environmental Protection Agency to dramatically weaken the wording of its public statements about the air quality in New York City in the days after the September 11, 2001 attacks on the World Trade Center."[62] As Jim Hightower has noted, as a presidential advisor Connaughton "has led the charge to weaken the standards of getting arsenic out of our drinking water, and he has steadily advised Bush to ignore, divert, stall, dismiss, and otherwise block out all calls for action against the industrial causes of global warming."[63]

As an attorney, Connaughton specialized in defending big corporations charged with violating Superfund laws (i.e., toxic waste clean-up and disposal regulations). Among his clients were General Electric, Atlantic Richfield, the Chemical Manufacturer's Association, and the American Smelting and Refining Company (ASARCO). In 1993, Connaughton co-authored an article for a law journal entitled, "Defending Charges Of Environmental Crime—The Growth Industry Of The 90s."[64] Indeed, he wrote from what he knew.

In fact, one of his firms' clients, General Electric, is said to be responsible for more Superfund sites than any other company in the nation.[65]

As a lobbyist Connaughton has lobbied on environmental issues, specifically those relating to Superfunds, on behalf of major corporate interests including the Aluminum Company of America, ASARCO, Atlantic Richfield, and the Chemical Manufacturers.[66] At the time of this writing, ARASCO, a miner, smelter, and refiner of copper and molybdenum, has filed for bankruptcy protection in Texas. Its liabilities for environmental and asbestos related clean-up "exceed its assets by $400 million."[67] Connaughton defended President Bush's refusal to sign the Kyoto treaty by saying that it would have cost America "nearly 5 million jobs," even though the administration's own Department of Energy strongly refuted the claim.

In Congress, the forces against environmental protections are strong—as strong as their ties to fundamentalist Christianity. For example:

Senate Majority Leader BILL FRIST (R-TN) voted repeatedly for increased construction of roads in national forests, supported Gail Norton for Secretary of the Interior, rejected the Superfund tax to clean up sites contaminated with hazardous chemicals, and had an extremely poor rating from the League of Conservation Voters in 2004. Meanwhile, Frist's ties to the fundamentalist movement have led Christian groups such as evangelsociety.org to name him as their top candidate for the Republican presidential candidate in 2008, and a *New York Times* editorial to declare, "Right-wing Christian groups and the Republican politicians they bankroll have done much since the last election to impose their particular religious views on all Americans. But nothing comes close to the shameful declaration of religious war by Bill Frist, the Senate majority leader...."[68]

JAMES INHOFE (R-OK), the chair of the Senate Environment and Public Works Committee, is "the Senate's most outspoken environmental critic," says *Grist* magazine journalist Glenn Scherer, and "is also unwavering in his wish to remake America as a Christian state."[69] A regular speaker to the Christian Coalition, Inhofe was one of the authors of President Bush's "Clear Skies Proposal."* (He also made the news when he requested the tax records of two non-profit organizations that

---

* The "Clear Skies" initiative would ease regulatory controls on utilities and set up a broader market-based system to reduce smokestack emissions.

criticized the proposal.[70]) In one notable speech from the floor of the Senate, Inhofe defended arsenic in drinking water, castigated UN weapons inspector Hans Blix for comments about climate change, and proclaimed, "Global warming is the greatest hoax ever perpetrated on the American people."[71]

House Majority Leader TOM DELAY (R-TX) has long sought the repeal of the Clean Air Act, referring to the Environmental Protection Agency as "the Gestapo of government;" he has fought to cut their budget, and consistently and vigorously opposed environmental protections. DeLay's ties to extraction lobbies are so strong that he actually allowed business lobbyists from the American Petroleum Institute to sit down with committee staff and help draft legislation.[72] The *New York Times* quoted DeLay as saying that he was on a mission from God to promote a "Biblical worldview."

Speaker of the House DENNIS HASTERT (R-IL), a strong supporter of the Bush/Cheney energy plan, which excluded environmentalist advisors, is pushing hard for granting the timber industry greater access to National Forests, to open up ANWAR for drilling, and for rules that would ease auto emissions standards.

Majority Whip ROY BLUNT (R-MO), who sits on the Energy and Commerce Committee, voted to accelerate forest thinning projects, pushed to reduce corporate liability for hazardous waste clean up and has a 0% rating on environment from the League of Conservation Voters.[73]

Assistant Majority Leader MITCH MCCONNELL (R-KY), has consistently voted against pro-environmental legislation and worked to open up public lands and wilderness areas to more roads. He has a 0% rating from the League of Conservation Voters. He received funds from the Coal Lobby, and fought for tax relief for the Coal Industry. He serves on the Committee for Agriculture, Nutrition, and Forestry. He has a 100% approval rating from Christian Coalition, with whom he works closely. He arranged for the group to hold workshops in the U.S. Senate auditorium, giving the Coalition convenient access to U. S. Senators.[74]

Senate Republican Conference Chair RICK SANTORUM (R-PA) voted for opening up roads in national forests and blocked enforcing pollution rules for corporate environmental law violators. His voting record has earned him a 0% approval rating from the League of Conservation Voters, and a 100% approval rating from the Christian Coalition. Santorum introduced legislation opposing the teaching of evolution in the schools, and was named by *Time* magazine as one of America's "25 Most Influential Evangelicals" and the *New York Times* called him "the nation's pre-eminent faith-based politician."[75]

With so many individuals in the Bush administration and the Republican Party pushing so hard against protecting the environment, the question begs, *is this part of their Christian directive*? Do such politicians believe that the earth has an endless supply of resources, or, at the very least, sufficient resources to last until Armageddon? Or is it that these politicians are conveniently being advised by reconstructionists on the Christian Right who believe in the "End of Times," and, as a consequence, supporting an American business culture that creates huge profits for companies at the expense of our very existence?

Divinity scholar S. R. Shearer identifies George W. Bush as being squarely in the camp that believes the latter of the two beliefs—that is, that Christ will return after a one thousand year Christian dominion over the earth. In his paper, "George Bush, The Promise Keepers, and the

Principle of Messianic Leadership," Shearer notes Bush's affiliation with the Promise Keepers as evidence. The Promise Keepers is a Dominionist men's group that describes itself on its website as "a Christ-centered organization dedicated to introducing men to Jesus Christ as their Savior and Lord; and then helping them to grow as Christians." Its motto: "Real Men Matter."[76] Women may not join (and indeed, among its tenets is an opposition to women in the ministry, the sort of thing that in 1997 led NOW president Patricia Ireland to label the group a "stealth male supremacist group").[77] Founded by the former head football coach of the University of Colorado, Bill McCartney, one of the group's principal leaders is Tony Evans, the pastor of the Oak Cliff Bible Fellowship in Dallas, Texas. Evans is a confidante and spiritual advisor to the President, has hosted White House prayer breakfasts, and regularly attends meetings with the President to consult on his "faith-based initiative."

As Shearer writes, "Most of the leaders of the Promise Keepers embrace a doctrine of 'End Time' (eschatology), known as 'dominionism.' Dominionism pictures the seizure of earthly (temporal) power by the 'people of God' as the only means through which the world can be rescued....

It is the eschatology that Bush has imbibed; an eschatology through which he has gradually (and easily) come to see himself as an agent of God who has been called by him to 'restore the earth to God's control.' "[78]

As for the character of the wealthy dominionists who contributed so much to the Bush campaign, Shearer says, "the 'new dominionists' think that the 'Good Life' is their due in the 'here and now' because they are 'sons and daughters of the Most High.' "

Meanwhile, Tony Evans has preached to his congregation that George W. Bush is the Messiah.[79] According to author Michael Ortiz Hill, "The Reverend Billy Graham taught [President] Bush to live in anticipation of the Second Coming but it was his friendship with Dr. Tony Evans that shaped Bush's political understanding of how to deport himself in an apocalyptic era. Dr. Evans...taught Bush about 'how the world should be seen from a divine viewpoint,' according to Dr. Martin Hawkins, Evans' assistant pastor."[80]

While no one claims that the President sees himself as the Messiah, evangelist author James Wallis, in a conversation with former long-time *Wall Street Journal* political reporter Ron Suskind, described Bush as a "Messianic American

Calvinist."[81] In an article in which Suskind depicts the Bush administration as an entirely "faith-based presidency," Wallis, of the Christian activist organization Sojourners, says "Faith can cut so many ways...but when it's designed to certify our righteousness, that can be a very dangerous thing." The article—in the *New York Times* Sunday Magazine—gained national attention not for its characterization of the President's religious views so much as for the remarks of one unidentified Bush aide, who told Suskind that most mainstream journalists suffered from a "reality-based" worldview. But the conservative "faith-based" Bush presidency, said the aide, reflected something other: "We're an Empire now, and when we act, we create our own reality." [82]

It's a worldview shared by others associated with the administration. Take Lt. Gen. William G. "Jerry" Boykin, who, in addition to being "assigned the task of tracking down and eliminating Osama bin Laden," and being part of the chain of command that oversaw prisoner care at Abu Ghraib prison in Iraq, is a deputy undersecretary of defense. As Richard T. Cooper reported in the *Los Angeles Times* on October 16, 2003, Boykin "is also an outspoken evangelical Christian who appeared in dress uniform and polished jump boots before a religious group in Oregon in June to declare that radical Islamists hated the United States 'because we're a Christian nation, because our foundation and our roots are Judeo-Christian...and the enemy is a guy named Satan.' Discussing the battle against a Muslim warlord in Somalia, Boykin told another audience, 'I knew my God was bigger than his. I knew that my God was a real God and his was an idol.' " On another occasion, Cooper reported, Boykin shared an Apocalyptic vision with an audience, telling them "We in the army of God, in the house of God, kingdom of God have been raised for such a time as this." He told another group of George Bush, "He's in the White House because God put him there."[83] He amplified this before another group, saying of George Bush, "Why is this man in the White House? The majority of Americans did not vote for him. Why is he there? And I tell you this morning that he's in the White House because God put him there for a time such as this."[84]

Nor have close associates of the administration been shy about their public comments concerning the fundamentalist leanings of the Republican government. House majority leader Tom DeLay, for example, told a church group that he was on a "mission from God to promote a 'biblical worldview,' in American politics," according to a December 18, 2002

New York Times column by Paul Krugman. Said DeLay: "Only Christianity offers a way to live in response to the realities that we find in this world—only Christianity." [85]

And of course, George W. Bush himself does not deny that he believes this. As he told reporters on Air Force One on the flight home from the funeral of Pope John Paul II, "My faith is strong. The Bible talks about, you've got to constantly stay in touch with the Word of God in order to help you on the walk [through life]."[86] As for whether that faith instills a belief that we are in the End of Times, Rick Perlstein of the *Village Voice* reported that Bush consulted with Apocalyptic Christians as he crafted negotiations with Israel over the Gaza Strip.[87]

However Bush thinks of himself or the orientation of his administration, at least one of his supporters adamantly refutes the notion that Bush is the Messiah—because he believes himself to be the Messiah. The Reverend Sun Myung Moon, founder of the Unification Church and owner of the vehemently pro-Bush *Washington Times* newspaper, has organized many dominion evangelicals in support of Bush. Beyond his financial support to dozens of far-right organizations, Moon has also, as Marc Fisher and Jeff Leen reported in a front page *Washington Post* story on November 23, 1997, donated $3.5 million to save Jerry Falwell's foundering Christian school, Liberty University, and "sponsored a series of speeches by George and Barbara Bush in Asia and the United States, with total fees estimated at about $1 million."[88]

More recently, his support of far-right politicians apparently earned Moon an unusual perk: the right to hold a ceremony in the Dirksen Senate Office building in Washington, and to have numerous members of Congress, the Senate, and the Bush administration attend. In a "coronation-like ceremony" arranged by the office of Republican Senator John Warner, bejeweled crowns were placed on the heads of Moon and his wife by Democratic congressman Danny Davis of Illinois, and an "Ambassador of Peace Medal" was bestowed upon Moon by Republican congressman Roscoe Bartlett of Maryland.[89] In a lengthy speech, Moon, who served an 18-month prison sentence in a Federal penitentiary for tax evasion and obstruction of justice in 1982, declared himself mankind's "savior, Messiah, Returning Lord and True Parent." Also in attendance, according to the Congressional newspaper *The Hill*: Republican Senators Lindsey Graham of South Carolina and

Mark Dayton of Minnesota, Republican Congressman Curt Weldon of Pennsylvania, and top Republican strategist and White House adviser Charlie Black.[90]

Of course, in a world of extremists, the pendulum swings both ways: An internet search quickly reveals a passionate discussion underway amongst fundamentalists seriously considering the idea that the president is the "beast" foretold in the Book of Revelation, and turns up sites such as www.bushistheantichrist.com detailing the theory.

But the search also yields the notion that we live in a time when ideas formerly considered extreme are held by people not usually suspected of holding them. Thus, an oft-linked news story by Wayne Madsen, a Washington, D.C.-based investigative journalist and former intelligence officer with the National Security Agency, who reported on the Counterpunch news site that toward the end of his life, Pope John Paul II, according to sources close to the pontiff, wished "he was younger and in better health to confront the possibility that Bush may represent the person prophesied in Revelations." Madsen wrote, "The Pope worked tirelessly to convince leaders of nations on the UN Security Council to oppose Bush's war resolution on Iraq. Vatican sources claim they had not seen the Pope more animated and determined since he fell ill to Parkinson's disease."

As Madsen noted, even though a non-secular leader, in at least one of his public statements the Pope's vision of impending Apocalypse sounded more realistic than the one he supposedly feared was being led by George Bush: "We are now standing in the face of the greatest historical confrontation humanity has gone through. I do not think that wide circles of the American society or wide circles of the Christian community realize this fully. We are now facing the final confrontation between the Church and the anti-Church, of the Gospel versus the anti-Gospel."[91]

CHAPTER 5

THE OTHER
FUNDAMENTALISTS

Analyzing the Wise Use strategies of an industry-funded "grass-roots" activism—known to critics as "astro-turf" activism—and of highjacked language, David Helvarg in *The War Against the Greens* writes, "One of Wise Use's major contributions to politics has been its deliberate distortion of language, the adaptation of green-sounding names as industry camouflage: the Environmental Conservation Organization (wetlands developers), Concerned Alaskans for Resources and the Environment (the timber industry), and the Greening Earth Society (coal-fired utilities arguing the benefits of global warming). Today, the administration's anti-environmental legislation goes by names like 'Healthy Forests' and 'Clear Skies,' while a combination of industry lobbyists and true believers argue that the same unfettered markets that gave us choking smog and burning rivers in the past have been transformed into the 'new environmentalism' of the 21st century. I recently read a government energy report that talked of 'the sustainable use of non-renewable resources,' and attended a Boston conference where I met consultants who help developers deal with 'environmentally challenged sites.' Translation: wild places containing rare and endangered species or legally protected wetlands."[92]

The fundamentalist Right has increased the number of think tanks that generate this kind of doublespeak and develop and disseminate anti-environmental propaganda. Usually, these think tanks also reveal how entwined the political right is with dominionists, reconstructionists, and other right-wing religious extremists.

The Acton Institute, for example, is a think tank co-founded by evangelist James Dobson to help set the political agenda for the New Christian Right. The chairman of the Acton Institute David Humphreys is the owner of Tamko Asphalt Products, and its board consists of individuals from oil companies, such as Sidney J. Jansma, Jr. of Wolverine Gas and Oil, and James L. Johnston of Amoco Corporation; and others from right-wing think tanks, such as Doug Bandow of the Cato Institute, Michael Novak of the American Enterprise Institute, and Jennifer Roback Morse of the Hoover Institute (famous for a membership including Ed Meese, Dinessh D'Souza, and Condoleeza Rice). The Acton Institute is in the business of turning out information for the New Christian Right's corporate and governmental interests to justify their anti-environmental policies. For example, as recipients of funding from Exxon/Mobil, they have dutifully generated misinformation regarding global warming, "debunking" the work of scientists that link emissions pollution and climate change.[93] A long essay on the Acton website entitled "The Catholic Church and Stewardship of Creation," for instance, declares that global warming "remains speculative and based on incomplete computer models rather than on demonstrated science," and "might cost man and nature a great deal if we rush to impose dramatic limits on fossil-fuel use in a misguided attempt to solve a problem that may not even exist."[94]

More famously, the Acton Institute developed its "Cornwall Declaration," which documents "the principles and aspirations of creation-based Environmental Stewardship." One section clarifies that, "Some unfounded or undue concerns [of environmentalists] include fears of destructive manmade global warming, overpopulation, and rampant species loss." Another says, "Public policies to combat exaggerated risks can dangerously delay or reverse the economic development necessary to improve not only human life but also human stewardship of the environment."[95]

In "Dominion *and* Stewardship: Believers and the Environment," an April 2004 "Acton Institute Commentary," Samuel Gregg, director of The Academic Research Center of the Acton Institute, takes the occasion of Earth Day to attack Christians who support a protective environmental agenda—out of fear, he explains, "that the framework through which Christians view the environment" may "slip into the wilderness of a type of neo-pantheism." Gregg attacks respected environmentalists such as Peter Singer ("more known for his advocacy of euthanasia and infanticide"), Paul Ehrlich (who he implies is an advocate of genocide, whose "claims have legitimized in advance various 'depopulation programs' "), and Leonardo Boff (a practitioner of "junk science"), all from an arch-Christian viewpoint. Interest in Earth Day has faded, he says, owing to "the fact that the dire consequences predicted by many ecologists have not eventuated." So, asks Gregg, "How, then, should believers think about events like Earth day?" His answer: with the "classic criteria... provided by the Book of Genesis"—in other words, through dominion theology: "The idea of dominion encapsulates the notion that human beings exercise a unique place in God's created order. They alone are charged with authority over the material world, and the responsibility of exercising it in ways that allow God's original Creative Act to be further unfolded. In this sense, human beings are co-creators."[96]

For Earth Day 2005, Gregg, whose Acton biographical note identifies him as "moral philosopher," again derides environmentally-minded Christians and says environmentalism, after a series of "neo-pagan tendencies that emerged throughout the 1990s," is down to "a dwindling number of liberation theologians, ex-priests, and self-described feminist eco-theologians."[97] The solution to any remaining environmental problems, he says, is to simply remove governmental regulation:

> One also searches in vain through the writings of many Christian environmentalists for recognition of the strength of private property-based solutions to environmental problems.

The beauty of these solutions does not just lie in their proven effectiveness. Utility and effectiveness have their place in the Christian moral life, but they are strictly subordinate to the demand to avoid evil and to participate in the moral goods that are at the root of human flourishing.

From this standpoint, the moral strength of private-property solutions is to be found in the manner in which they place direct responsibility for the natural world squarely in the hands of real flesh-and-blood individuals. To this extent, they also prevent us from shirking such responsibilities by delegating them to politicians and bureaucrats.[98]

Fred Krueger is also a devout Christian—he calls himself a "traditional Christian"—who, for many years, has dedicated himself to teaching the opposite of positions such as those of Gregg and the Acton Institute. Krueger is head of the Religious Campaign for Forest Conservation, a coalition of churches, synagogues, and other religious organizations working on forest conservation and wilderness issues, and against the anti-environmental policies of the Bush administration. "The Bible and all Biblical scholars of note demand that Orthodox Christian theology directs stewardship of the earth," he explained to me.

Few men know the doublespeak strategies of spin better than Fred Krueger, who worked as a campaign coordinator for the Republican National Committee on college campaigns. "There is very little integrity in that business," Krueger told me. "Anything goes to get the vote, to gain power for the party."

His own experiences with the Bush White House convinced him that the administration's position was essentially utilitarian and not a genuine religious ideology. "There is," he told me, "a conflict between the forces of greed in government and what the Bible says about having stewardship of the earth—really protecting the earth that God has made. God is present in all things. There is no compatibility with the utilitarian or profit making worldview and the religious worldview, because the religious worldview sees intrinsic value in all things, because they are created by God, and God in fact is present within all things."

Krueger said he and other religious leaders from all denominations met repeatedly with representatives from the Bush administration to advocate for protections for the nation's forests, for global climate concerns, and for the protection of endangered species. As Krueger relates it, the first meeting was with the heads of several departments within the administration, including James Connaughton, the head of the Council of Environmental Quality, and others from the Office of Regulatory Affairs. They also met with David Quo, then deputy director the Office of Community and Faith based Initiatives. Krueger told me, "One of our researchers told us that Quo had formerly worked for one of the government intelligence agencies, and also with Ralph Reed's Century Strategies, a political consulting firm."[99]

However, Krueger said, "These people weren't interested in what we had to say. In the last meeting we attended, they sent a young man who was so low in importance, it became very clear to us how low in priority we were, a group of ministers and other religious people, to the Bush administration."

It was clear to Krueger that corporate profit and politics dominated the current government's perspective on environmental protections. The Bush administration, he told me, has skewed the Bible to support bad capitalism. And capitalism without the moral restraint of Christianity, he said, "becomes cruel and despotic." Krueger maintains that Christianity dictates that the economy is subordinate to the environment, not the other way around. As he writes on The Religious Campaign for Forest Conservation website:

As people of faith, we find intrinsic value in all parts of creation. We find that the spiritual qualities and particularly the non-economic values of forests soar far beyond the economic and other quantitative measurements which have been applied to assess forest worth. These measurements of forest value are clearly inadequate. For reasons rooted in the moral principles of Judaism and Christianity, for reasons of sound science untainted by commercial manipulation, and for reasons dealing with our human responsibility to provide a healthy world for future generations, we see ourselves as morally bound to work for the protection and preservation of the forests and wilderness as crucial elements of the planet's life support system. We do this as our obedience to God and our service to humanity and particularly the children of today and tomorrow.[100]

Krueger says that he takes religious leaders on Wilderness retreats to "gain the experience of having a relationship with God," because he believes in the power of nature to spiritually transform Christian leaders. He cited David Brower, one of the founding fathers of the environmental movement, who said only religious institutions "could turn the tide of environmental destruction."[101]

Another influential figure in the Christian pro-environment movement is Peter Illyn, an evangelical Christian and executive director of the Restoring Eden Ministry—a network of environmentally-minded Christians, as its website puts it, "interested in combining their faith and their concern for nature," aiming to " 'speak out for those who cannot speak for themselves' (Proverbs 31:8) as advocates for the wild habitat, native species and indigenous cultures of the Arctic." [102]

When I first contacted him, Illyn was leading a group of college students through Washington, DC to lobby members of Congress to vote against opening up the arctic wilderness for oil drilling. Afterwards, I caught up with him on his cell phone while he was on a tour of the South. "I'm driving through the Bible Belt," he told me. "I'm speaking to Southern Baptist Conference colleges about stewardship of the earth, getting the young people on board."

Illyn also works with spiritual leaders of various indigenous communities and tribal Christians as well as traditional evangelical Christians, such as those of the Southern Baptist Conference. He disagrees with New Christian Right evangelists such as Tim LaHaye and John Alquist, who claim that God is not in Nature, that the earth is cursed, or that the earth may have to perish before Christ will return. To Illyn, the dictate lies in Revelations 11:18: "The nations were angry; and your wrath has come. The time has come for judging the dead, and for rewarding your servants the prophets and your saints and those who reverence your name, both small and great—and for destroying those who destroy the earth." In other words, says Illyn, "Those who destroy the earth, well, I wouldn't want to be in their shoes! God will destroy the people who are destroying the earth."

In a 1997 interview with *High Country News* journalist Jeffrey Smith, Illyn recounted his epiphany about the divine command to protect the environment. "Illyn was a Pentecostal minister at a Foursquare church in Portland," writes Smith, "when, in 1989, he went on sabbatical and spent four months hiking 1,000 miles of the Pacific Coast Trail with a pack of llamas...." Illyn told Smith, "I went into the woods a Pentecostal preacher, and I saw the clearcuts, and I saw the pollution, and I saw the devastation, and I came home an evangelical environmentalist."[103]

In my own conversation with Illyn, he spoke of why so many people believe that we are in the "End of Times" now, and of how they might be influenced by those who use this philosophy to justify environmental destruction. On speaking tours like the one he was on now, he told me, people are constantly asking him questions fundamental to religion, such as, "Where is the earth from? Where is Humanity in this? What is the future? What is Eternity?"

"I'm a creationist," Illyn told me, "but I think we have to introduce what I call a new 'meta-level narrative' to Christians. You know, people think of miracles as being unusual. The truth is, miracles are the Laws of Nature. When you see how plants transform carbon dioxide, and other amazing processes made by God when He created the earth—those are miracles."

As to his understanding of the Book of Revelation, Ilyn says, "I read that to mean that there may be a cleansing of corruption, and there may be some scars, but it is clear that the Earth shall be renewed and Christ will return to a restored earth." As to whether dominionists are using a different interpretation of Revelations to justify the purposeful implementation of environmentally harmful policies, he said. "I don't see a focused agenda from them to destroy the Earth for Christ's return. They might be interpreting the Bible and using the dominion message to justify their actions that are based on greed. It's really all about free market capitalism and greed."

Peter Illyn and Fred Krueger are not alone in promoting the protection of the environment in the evangelical world. In a March 2005 *New York Times* article, "Evangelical Leaders Swing Influence Behind Effort to Combat Global Warming," Laurie Goodstein reported that the National Association of Evangelicals (NAE), an umbrella group of 51 church denominations, had adopted a platform called "For the Health of the Nation: An Evangelical Call to Civic Responsibility," which included a plank on "creation care":

> Because clean air, pure water, and adequate resources are crucial to public health and civic order, government has an obligation to protect its citizens from the effects of environmental degradation.... We must therefore approach our stewardship of creation with humility and caution. Human beings have responsibility for creation in a variety of ways. We urge Christians to shape their personal lives in creation-friendly ways: practicing effective recycling, conserving resources, and experiencing the joy of contact with nature. We urge government to encourage fuel efficiency, reduce pollution, encourage sustainable use of natural resources, and provide for the proper care of wildlife and their natural habitats.[104]

This statement was seen by many observers of the New Christian Right as well as pro-environment evangelists as an important departure in evangelical Christians' relationship to environmental issues. The NAE, a major evangelical group that has long looked askance at environmentalism, was embracing the idea of environmental stewardship as part of their covenant with God—and that has now, apparently, carried over into an activist political stance on the environment.

The Rev. Ted Haggard, president of the NAE, told the *Times*, "The question is, will evangelicals make a difference, and the answer is, the Senate thinks so. We do represent 30 million people, and we can mobilize them if we have to." [105]

The NAE's vice president, Reverend Rich Cizik, says his change of heart came when he was "dragged" to a conference on climate change in Oxford, England in 2002, by Jim Ball, director of the Evangelical Environmental Network and founder of the "What Would Jesus Drive" campaign.* There he heard Sir John Houghton, co-chair of the Intergovernmental Panel on Climate Change and

* Gaining considerable attention upon its launch, when it was often characterized in the press as a Christian initiative against SUVs, the continuing effort, sponsored by the EEN, describes itself as follows on its website: "The purpose of the What Would Jesus Drive? educational campaign is to help Christians and others understand that transportation choices are moral choices, and to reflect upon the problems associated with transportation from a biblically orthodox, Christ-centered perspective." See: www.whatwouldjesusdrive.org.

former professor of atmospheric physics at Oxford University, speak about the Christian responsibility to protect the earth. As Goodstein detailed in the *Times*: "Cizik said he had a 'conversion' on climate change so profound in Oxford that he likened it to an 'altar call,' when nonbelievers accept Jesus as their savior."[106] Cizik and EEN director Ball asked Houghton to speak to a group of Christian evangelicals in Washington, DC.

In his subsequent speech, Houghton detailed the scientific proof for global warming, including research on the burning of fossil fuels, and implored the U.S. to immediately take measures to drastically reduce emissions. Houghton proclaimed:

> I believe God is committed to his creation. He demonstrated this most eloquently by sending his son Jesus to be part of creation and by giving to us the responsibility of being good stewards of creation. What is more I believe that we do not do this on our own but in partnership with him—a partnership that is presented so beautifully in the early chapters of Genesis where we read that God walked with Adam and Eve in the garden in the cool of the day.[107]

Ball, Cizik, and the NAE have come under attack by many in the New Christian Right since inviting Houghton to speak—most notably by Republican Senator James Inhofe of Oklahoma. Inhofe, an evangelical Christian, chairs the Senate Committee on Environment and Public Works and famously declared from the Senate floor in Spring 2003 that global warming is a "hoax."[108] (In that same speech, Inhofe declared that, "The Kyoto Protocol has no environmental benefits; natural variability, not fossil fuel emissions, is the over-whelming factor influencing climate change."[109] Perhaps not coincidentally, Inhofe is second only to fellow Republican Senator John Cornyn of Texas as the recipient of campaign contributions from the oil and gas industries.)

I reached Cizik by phone on a busy morning to ask him about the controversy. A relaxed, friendly man, who seems eminently believable when he says he has made many friends in Washington during his twenty-six years there, he shrugs off Inhofe's attack, saying, "I have a thick skin." Cizik says he doesn't believe that Inhofe's behavior towards environmental issues comes from a Biblical or apocalyptic directive. "It's more driven by politics," he says. "What comes first? The politicians are rewarded for their anti-environ-ment views with funding from special interests who oppose environmental protections. I don't believe that they first receive the money and then go on to represent the special interests views."

I asked Cizik if evangelical Christians can be environmentalists too. "Absolutely!" he replied. But he is careful not to refer to himself an environmentalist. He prefers using the term "creation care" to talk about environmental issues. As he explained in a *New York Times* interview, "Environmentalists have a bad reputation among evangelical Christians. They keep kooky religious company.... Some environmentalists are pantheists who believe creation itself is holy, not the Creator."[110]

In the end, it remains to be seen how evangelicals such as Krueger, Illyn, Ball, or Cizik will effect the agenda within the evangelical Christian community and among the environmental policymakers. So far, however, they seem undaunted by the vehement opposition such as that led by Inhofe. As Cizik told the *New York Times* "I don't think God is going to ask us how he created the earth, but he will ask us what we did with what he created."[111]

ENDNOTES

1   *A Wolf in the Garden: The Land Rights Movement and the New Environmental Debate,* edited by Philip D. Brick and R. McGreggor Cawley (Lanham, Maryland: Rowman & Littlefield Publishers, 1996), available at www.cdfe.org/wiseuse.htm.

2   Ron Arnold, "Defeating Environmentalism," *Logging Management Magazine,* April 1980, pp. 40-41, as quoted in Tarso Ramos, "Extremists and the Anti-Environmental Lobby: Activities Since Oklahoma City," a research paper for the Western States Center, available at www.westernstatescenter.org/archive/wupep/extreme.html.

3   Claude Emery, "Share Groups in British Columbia," Library of Parliament Research Branch, 10 December 1991, p. 12., as quoted in Ramos.

4   David Helvarg, "Wise Use in the White House," *Sierra Magazine,* September 2004, available at www.sierraclub.org/sierra/200409/wiseuse.asp.

5   Helvarg, "Wise Use in the White House."

6   Helvarg, "Wise Use in the White House."

7   First Arnold quote is from "Fund Raisers Tap Anti-Environmentalism," *The New York Times,* December 19, 1991; Second citation is from a CNN interview with Ron Arnold, May 30, 1993.

8   Helvarg, "Wise Use in the White House."

9   "Anti-gay Battlefield in Oregon," Associated Press, June 11, 1992.

10 Tom Kenworthy, "Both Parties Shy Away From Antigay Rerun; Oregon Conservative Group Shows Resilience," *The Washington Post,* October 26, 1994.

11 Vanessa Ho, "Aid Sought to Fight Anti-Gay Group," *Seattle Times,* March 30, 1992.

12 Dave Mazza, *God, Land and Politics: The Wise Use and Christian Right Connection in 1992 Oregon Politics* (Portland: Western States Center and Montana AFL/CIO, 1993).

13 Charles P. Alexander, "Gunning for the Greens," *Time Magazine*, February 3, 1992.

14 Helvarg, "Wise Use in the White House."

15 "Talking with Mark Crispin Miller, Author of *Cruel and Unusual: Bush/Cheney's New World Order*," Buzzflash.com, July 23, 2004; available at www.buzzflash.com/interviews/04/07/int04037.html.

16 See Matthew J. Slick's estimates at The Christian Apologetics and Research Ministry, www.carm.org/bible/biblewhen.htm.

17 See www.creationcare.org/resources/scripture.php.

18 Citation is from the Evangelical Environmental Network & *Creation Care* magazine website.

19 Joe Bageant, "The Covert Kingdom: Thy Will be Done, On Earth as It is in Texas," Counterpunch, May 25, 2004, available at counterpunch.org/bageant05252004.html.

20 Glenn Scherer, "The Godly Must Be Crazy: Christian-right Views are Swaying Politicians and Threatening the Environment," *Grist*, October 27, 2004. Available at www.grist.org/news/maindish/2004/10/27/scherer-christian/.

21 Scherer, "The Godly Must Be Crazy."

22 Austin Miles, *Setting the Captives Free: Victims of the Church Tell Their Stories* (Amherst, New York: Prometheus Books, 1990), p. 229.

23 See the National Council of Churches statement at www.ncccusa.org/news/14.02.05theologicalstatement.html.

24 Jeffrey Smith, "Christian Evangelicals Preach a Green Gospel," *High Country News*, April 28, 1997; available at www.hcn.org/servlets/hcn.URLRemapper/1997/apr28/dir/Feature_Christian.html.

25 Walter Olson, "Invitation to Stoning: Getting Cozy with Theocrats—Christian Reconstructionists," *Reason*, November 1998.

26 Karen Armstrong, *The Battle for God* (New York: Knopf, 2000), pp. 361-362.

27 Chip Berlet and Matthew Lyons, *Right-Wing Populism in America: Too Close for Comfort* (New York: The Guilford Press, 2000), p. 249.

28 See NCA statement, as cited above.

29 Scherer, "The Godly Must Be Crazy."

30 See www.allianceforamerica.org/Bios_Officers/Comstock_Bio.htm.

31 As David Helvarg noted in his book *The War Against the Greens* "One of Wise Use's major contributions to politics has been its deliberate distortion of language, the adaptation of green-sounding names as industry camouflage...."

32 Thomas Harding, "Mocking the Turtle: Backlash Against Environmental Movement" *New Statesman & Society*, September 24, 1993, p. 45.

33 Bruce Alpert, "Alliance For America Accused of Being A Front," *Times-Picayune*, September 25, 1993.

34 Timothy Egan, "Fund-Raisers Target 'perfect bogeyman'; Environmentalists Feeling Backlash," *New York Times*, December 19, 1991.

35 Alpert, "Alliance For America Accused of Being A Front."

36 Timothy Egan, "Fund-Raisers Target 'perfect bogeyman.' "

37 "Was Their Money Wisely Used?" *The National Journal,* July 8, 1995.

38 Alpert, "Alliance For America Accused of Being A Front."

39 Betty Mitchell, "Fisheries Official's Involvement in National Alliance Questioned," *The Virginian-Pilot,* May 24, 1995.

40 The Caples Crest Kirkwood Mountain Resort on Caples Mountain will pay nothing to the Federal government until the ski lifts are operational, which is expected about two to three years after completion of the restaurant. Once the lifts are operational, the Federal government will then receive 2% of revenues from lift tickets. Kirkwood will pay no local, state, or federal taxes. Source: Tim Cohee, President of Kirkwood Resort, telephone interview, July 7, 2005.

41 See www.mykirkwoodhome.com.

42 "Sierra Club v. Morton 1968," accessed at caselaw.lp.findlaw.com/scripts/get-case.pl?court=US&vol=405&invol=727.

43 Army Corp of Engineers, Archive of Memorandums of Understanding (MOU's) and Agreement (MOA's), available at corpslakes.usace.army.mil/employees/cec-won/mou-archive.html; see also a copy of the 1995 "Memorandum of Understanding" at www.wildwilderness.org/docs/95-mou.htm.

44 Department of the Interior press release, September 21, 2004, available at www.doi.gov/news/040921a.

45 Helvarg, "Wise Use in the White House."

46 See: https://host186.ipowerweb.com/~kenneth1/about.htm

47 Robert Dreyfus, "Reverend Doomsday," *Rolling Stone,* January 28, 2004.

48 Dreyfus, "Reverend Doomsday."

49 Jim Alquist and the Tell It Well, First Internet Christian Church. Accessed April 16, 2005 at www.tell-it-well.com/page38.htm.

50 See www.tell-it-well.com/page27.htm.

51 Among others: James Bell, "Problems with the Iraqi Wetlands and Their Reclamation," available at www.jameswbell.com/a013marshreclamation.html; Chaterjee, Pratap "Bechtel's Friends in High Places," Corpwatch, April 24, 2003, available at http://corpwatch.org/article.php?id=6548; Jim Vallette, Steve Kretzmann, Daphnie Whysham, "Crude Vision: How Oil Interests Obscured U.S. Government Focus On Chemical Weapons Use by Saddam Hussein," Sustainable Energy and Economy Network, March 2003, available at www.seen.org/pages/reports.shtml; The quote is from Bell.

52 Tim LaHaye, *Prophecy Study Bible* (AMG Publishers, 2000), p. 1285.

53 Mark Matthews, "Christians Preach Environmental Gospel." *High Country News,* February 19, 1996, available at www.hcn.org/servlets/hcn.Article?article_id=1650.

54 Jeremy Leggett, *The Carbon Wars* (New York: Routledge, 2001), p. 174.

55 Jennifer Hattam, "Lay of the Land," *Sierra Magazine,* May/June 2001, available at www.sierraclub.org/sierra/200105/lol1.asp.

56 See statement by Kalee Kreider, Environmental Media Services, March 10, 2005, Kalee Kreider www.bushgreenwatch.org.

57 Helvarg, David. "Wise Use in the White House."

58 See the work of Steven Weiss, The Center for Responsive Politics and OpenSecrets.org, www.opensecrets.org.

59 Ralph G. Neas, "The Federalist Society: From Obscurity to Power," People for the American Way, September 17, 2003. See pfaw.org.

60 John Nichols, "Secretary of Agribusiness," *The Nation,* December 7, 2004.

61 William Branigin and Jim Vandehei, "Johanns Nominated for Agriculture Secretary," *The Washington Post,* December 2, 2004, available at www.washingtonpost.com/wp-dyn/articles/A28011-2004Dec2.html.

62 See www.sourcewatch.org/index.php?title=James_L._Connaughton.

63 Jim Hightower, *Let's Stop Beating around the Bush: More Political Subversion from Jim Hightower* (New York: Viking, 2004); online excerpt at "Dirty Dozen Sub-cabinet Policy Operatives Actually Run Government," *Utne,* June 2004, http://www.utne.com/web_special/web_specials_2004-06/articles/11232-1.html.

64 Thomas C. Green and James L. Connaughton, "Defending Charges Of Environmental Crime—The Growth Industry Of The 90s," *The Champion Magazine,* National Association of Criminal Defense Lawyers, April 1993.

65 Robert S. Devine, *Bush Versus the Environment* (New York: Anchor Books, New York, 2004), p.24

66 Devine, *Bush Versus the Environment.*

67 Susan Gordon, "Asarco Seeks Bankruptcy Protection," *The News Tribune,* August 11, 2005.

68 Keith Miller, "Bible Con Weekly III: Politics," The Evangel Society, March 4, 2004 and April 15, 2005. See www.evangelsociety.org/miller/bcw3.html.; See also: "Bill Frist's Religious War," *New York Times,* April 16, 2005.

69 Scherer, "The Godly Must be Crazy."

70 See the research of Kenneth Quinnell, "Right Wing Watch: Politics: James Inhofe, Tom DeLay," at www.quinnell.us/politics/rww/individuals/index17.html.

71 James Inhofe, "The Science of Climate Change," United States Senate, July 28, 2003 and March 20, 2005; available online at inhofe.senate.gov/pressreleases/climate.htm.

72 Robert Dreyfuss, "DeLay, Incorporated," *Texas Observer,* February 4, 2000.

73 Jesse Gordon, On the Issues, January 1, 2000 and August 8, 2003, available at http://www.issues2000.org/House/Roy_Blunt_Environment.htm.

74 Nikhil Aziz, "Road to Victory 2004: Let's Take America Back...a hundred years," Public Eye, see www.publiceye.org/praccess/archive/v3n1/victory_aziz.html.

75 Michael Sokolove, "The Believer," *The New York Times*, May 22, 2005.

76 See www.promisekeepers.org.

77 See www.now.org/issues/right/promise/letter.html.

78 S.R. Shearer, "George Bush, The 'Promise Keepers,' And The Principles Of Messianic Leadership," November 28, 2001 and August 16, 2003, available at www.antipas-ministries.com/oldnews/george.html.

79 Michael Ortiz Hill, phone interview, March 2003.

80 Michael Ortiz Hill, "Bush's Armageddon Obsession, Revisited," *Counterpunch*, January 4, 2003, available at www.counterpunch.org/hill01042003.html.

81 However, in his autobiography, *A Charge to Keep*, published when he was still Governor of Texas, Bush wrote, "I could not be (governor) if I did not believe in a divine plan that supersedes all human plans." *A Charge to Keep* (New York: William Morrow, 1999), p. 6.

82 Ron Suskind, "Without a Doubt," *New York Times Magazine,* Oct 17, 2004.

83 Richard T. Cooper, "General Casts War in Religious Terms" *Los Angeles Times*, October 16, 2003.

84 See NBC News broadcast by Lisa Myers, October 15, 2003, http://www.msnbc.com/news/980764.asp?cp1=1

85 William Cook, "Armageddon Anxiety: Evil On the Way," *Counterpunch*. February 22, 2003, avalible at www.counterpunch.org/cook02222003.html.

86 George Bush, press briefing—Air Force One, en Route from Pope John Paul's funeral. April, 8 2005 www.whitehouse.gov/news/releases/ 2005/04/20050408.html.

87 Rick Perlstein, "The Jesus Landing Pad," *The Village Voice,* May 18, 2004, available at www.villagevoice.com/news/0420,perlstein,53582,1.html.

88 Marc Fisher and Jeff Leen, "A Church in Flux Is Flush With Cash," *The Washington Post*, November 23, 1997.

89 Charles Babington, "Warner Helped the Rev. Moon; Senator's Office Says He Arranged for Meeting Space in March," *The Washington Post*, July 21, 2004.

90 James Kirchick, "Lawmakers attend Moon 'coronation' in Dirksen," *The Hill*, June 22, 2004.

91 Wayne Madsen "Bush's 'Christian' Blood Cult: Concerns Raised by the Vatican," *Counterpunch*, April 22, 2003, see www.counterpunch.org/madsen04222003.html.

92 David Helvarg, *The War Against the Greens* (Johnson Books, 2004, revised edition).

93 Chris Mooney, "Some Like It Hot" *Mother Jones Magazine,* May/June 2005, p. 36.

94 "The Catholic Church and Stewardship of Creation," The Acton Institute, see www.acton.org/ppolicy/environment/theology/m_catholic.html.

95 "Cornwall Declaration," The Acton Institute, see www.acton.org/ppolicy/environment/cornwall.html.

96 Samuel Gregg, "Dominion and Stewardship: Believers and the Environment," Acton Institute, April 14, 2004, see www.acton.org/ppolicy/comment/article.php?id=193.

97 Samuel Gregg. "God, Man and the Environment," Acton Institute, April 20, 2005. see www.acton.org/ppolicy/comment/article.php?id=261.

98 Samuel Gregg. "God, Man and the Environment."

99 See www.censtrat.com/index.cfm?FuseAction=Home.Home.

100 Fred Krueger, "The Religious Campaign for Forest Conservation, see creationethics.org/index.cfm?fuseaction=webpage&page_id=1.

101 See Brower's essay, "The Sermon—A Healing Time on Earth" advocating "wilderness as church," www.wildnesswithin.com/heal.html.

102 Restoring Eden, www.restoringeden.org.

103 Jeffrey Smith, "Christian Evangelicals Preach a Green Gospel" *High Country News*, April 28, 1997, available at www.hcn.org/servlets/hcn.URLRemapper/1997/apr28/dir/Feature_Christian.html.

104 "For the Health of the Nation: An Evangelical Call to Civic Responsibility," The National Association of Evangelicals, see www.nae.net/index.cfm?FUSEACTION=editor.list&IDCategory=9.

105 Laurie Goodstein, "Evangelical Leaders Swing Influence Behind Effort to Combat Global Warming," *New York Times*, March 10, 2005.

106 Laurie Goodstein, "Evangelical Leaders Swing Influence Behind Effort to Combat Global Warming."

107 Sir John Houghton, "Global Warming—the Science, The Impacts, The Politics," a lecture at Trinity College, Cambridge, May 25, 2001, available at www.stedmunds.cam.ac.uk/cis/houghton/lecture5.html.

108 Senator James Inhofe, July 28, 2003; speech excerpts available at inhofe.senate.gov/pressapp/record.cfm?id=206907.

109 Inhofe.

110 Deborah Solomon, "Earthy Evangelist," *New York Times Magazine,* April 3. 2005.

111 Deborah Solomon, "Earthy Evangelist."

ACKNOWLEDGEMENTS

Many thanks to the many people who have helped me on this project and in this work: Erik Ferry, Harlan Stelmach, Dara Hellman, Tom Burke, Michael Ortiz Hill, Caroline Casey, Larry Bensky, Deepa Fernandes, and Mitch Jeserich, Jon Almeleh, Sonali Kohlhatkar, Kevin Danaher, Medea Benjamin, Julia Butterfly Hill. To my wonderful friends: Stephanie, Corey, Deborah Sue, Sondra, Chris Braun, Lori, Margie, and the Belrose gang. And all who I neglected while focused on this: to my beautiful family, Judy, Susie, Dick, Elizabeth, to my wonderful husband Kelly and magnificent daughters Becky and Sarah. I thank you all.